Simulation of nitrogen behaviour in soils

J. Beek and M. J. Frissel

UNIV. COLL. N.W.
BANGOR
LIBRARY

AF851073

pudoc

Wageningen
Centre for Agricultural Publishing and Documentation
1973

743223

ISBN 90 220 0440 6

© Centre for Agricultural Publishing and Documentation, Wageningen, 1973
No part of this book may be reproduced or published in any form, by print, photoprint, microfilm or any other means without written permission from the publisher.
Cover design: Pudoc, Wageningen

Printed in Belgium

Sci f2

Simulation of nitrogen behaviour in soils

Simulation Monographs

Simulation Monographs is a series on computer simulation in agriculture and its supporting sciences

Contents

This book describes the present state of development of a quantitative simulation model for the nitrogen behaviour in the soil.
This work has been carried out by a group of students and research workers in Wageningen, the Netherlands.

These are:

Agricultural University of Wageningen, Department of Theoretical Production Ecology:
J. Beek, F. van Dorp, W. M. Lindhoud, H. Stricker, C. T. de Wit

Institute for Soil Fertility, Haren (Groningen):
Th. J. Ferrari, G. J. Kolenbrander

Association Euratom–Institute for Atomic Sciences in Agriculture, Wageningen:
J. Beek, M. J. Frissel.

Address for comments and suggestions:
M. J. Frissel, P.O.Box 48, Wageningen, the Netherlands.

1 Introduction

UNIV. COLL. N.W.
BANGOR
LIBRARY

1.1 The problem

In the majority of soils most of the nitrogen is in organic compounds and only a small percentage is present in inorganic compounds. Depending on the type of decomposition process of the organic matter, organic nitrogen is mineralized into ammonia or inorganic nitrogen is immobilized by transforming it into organic nitrogen. The inorganic soil nitrogen is mainly present in the form of nitrate and ammonia. Ammonia may be transformed to nitrate by nitrification while nitrate may be transformed to volatile nitrogen compounds. All these processes are carried out by soil micro-organisms and consequently are influenced by temperature, moisture, pH and aeration of the soil. Since both the turnover rate between the organic, ammonium, nitrate and elementary form of nitrogen is very high and the influence of external factors like temperature and precipitation is considerable, it is hardly possible to estimate the inorganic nitrogen available for plants during a growth season, on the basis of chemical and physical experiments at one certain moment. An alternative approach of attaining a reliable estimate would be to use the information available in literature about the separate microbial processes. Then the problem is to integrate the knowledge about the separate processes into a larger model where all the involved processes are acting simultaneously. A unique opportunity to solve this problem is provided by the modern digital computer simulation techniques. The 'Continuous System Modeling Program' (CSMP/360), developed for the IBM 360 series of machines, has proved to be very suitable to simulate biological processes. Information about CSMP is given in the 'User's manual of the system /360 Continuous System Modeling Program'.

The purpose of this book is threefold, to present a summary of the work carried out so far, to promote an exchange of ideas about this approach and to inform soil microbiologists which kind of experiments have to be carried out to provide model builders with the data needed to develop and test such an approach. It is stressed that this description of the simulation model only represents an intermediate step in its development so that no definite conclusions can be drawn.

1.2 Simulated processes

The microbial decomposition of organic matter with its accompanying mineralization or immobilization, the production of humus and the nitrification process are defined quantitatively. This approach is not usual so that in the literature particular information is scarce.
Even for the simple soil systems which are simulated here (i.e. cultivated soils in aerobic conditions, with a high buffer capacity) calcu-

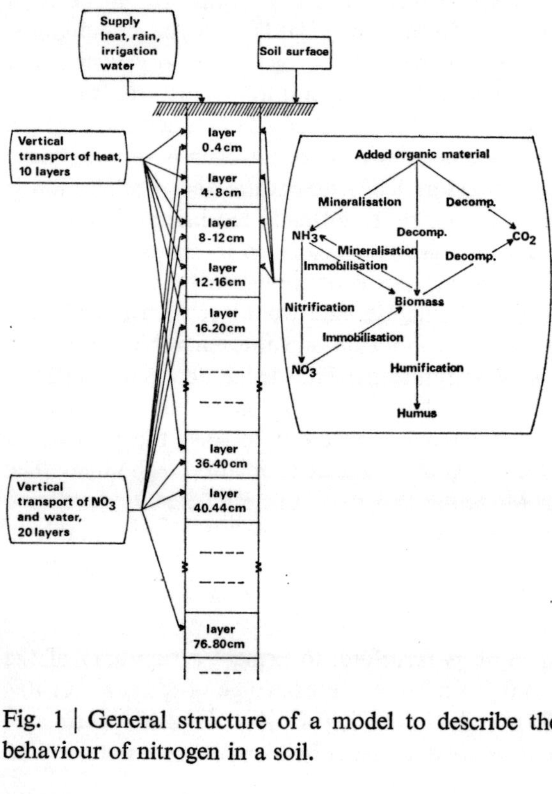

Fig. 1 | General structure of a model to describe the behaviour of nitrogen in a soil.

lated data according to the simulation have not been compared with reported results in the field because of lack of data. The decomposition processes are described in Chapter 2, the nitrification process in Chapter 3.
The physical processes of the movement of water, heat and salt included in this model have been copied, with only slight modifications, from existing models, for a comprehensive description of above mentioned work see de Wit & van Keulen (1972).
The simulation process of the movement of heat can be used to calculate the new temperature of the soil, brought about by a change in air temperature or by a flow of heat caused by the infiltration of irrigation water supplied at the surface. The water content of the soil, which depends on the amount of water supplied and its flow rate, may be calculated from the simulation of the transport of irrigation water in an unsaturated soil. Both the temperature and the water content of the soil influence the microbial processes. The program for the leaching of salt is used for the calculation of the amount of nitrate leached since this anion is hardly adsorbed by the soil complex. The transport of heat and water are briefly described in Chapter 4.

In due course the denitrification and fixation of air nitrogen by bacteria have to be included in the model. Secondly, provision has to be made for the influence of the soil pH on the microbial processes, but also the reverse, the change of the soil pH brought about by these processes has to be included.

1.3 A compartment simulation model

Figure 2 shows a soil column of finite length taken from an infinite slab in which the physical and microbial processes take place. To simulate these processes the column is divided into 20 equal compartments or layers with thickness TCOM (TCOM = 4 cm). Hence, it is possible to define the conditions and the processes taking place per compartment. In this particular model, decomposition and nitrification only occur in the upper four compartments, the movement of heat in the upper ten compartments and the transport of water and nitrate in all twenty compartments. The water content, temperature and composition of the soil solution are equal in all compartments at the start of the simulation. It is stressed, however, that these assumptions are

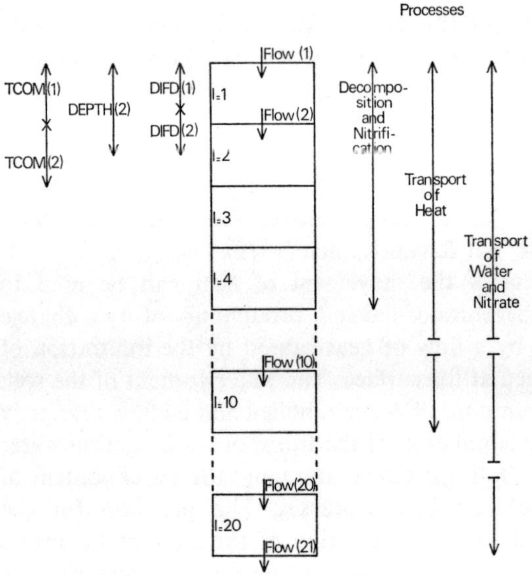

Thickness of compartment or layer TCOM = 4 cm
Distance between centers of layers DIFD
" " surface and center of a layer DEPTH
Area AREA = 1 cm^2

Fig. 2 | Geometry of the system and symbols, used to describe the transport process.

certainly not essential. The number of layers describing the various processes, layer thicknesses, differences in composition of the layers, differences in temperature can all be provided for within the frame of the developed model.

2 Decomposition processes

2.1 Basic processes

The designers of a simulation model must discriminate between essential and less essential aspects of the actual processes.
The micro-organisms in the soil are thought to decompose organic matter, forming free CO_2, new microbial tissue, humus, and to transform nitrogen. The ammount of CO_2 evolved is supposed to be a measure for the activity of the micro-organisms and so it is assumed that the amount and kind of organic material supplied, the moisture, the temperature and some other soil characteristics influence the rate of decomposition. It is known that the relative rate of decomposition of organic matter, such as wheat straw, decreases during decomposition. According to Kononova (1961) this decrease is because the easily decomposable components of the added material are used first while the more resistant ones remain. Hence in the model it is assumed that the organic matter supplied consists of four components, their relative amounts depending on the kind of matter added. These components are protein, sugars, cellulose and lignin which, in this order, have shown to be more resistant to microbial attack. The relative rate of decomposition of each component, however, is supposed to be independent of the fraction decomposed. Obviously the total simulated rate of decomposition of the added material has to be the same as the experimentally measured rate of decomposition. If the water content and temperature of the soil are not optimum, reduction factors have to be built in to account for the decrease in decomposition.

Usually organic matter of the soil is present as humus, less humified compounds originating from previous supplies of manure or old roots and stubble, and microbial tissue. All these organic components decompose but at different rates. Thus quantitative data is necessary on the presence of these components in the soil and their relative rates of decomposition. Since these data are very scarce, it is assumed in this

case that the soil, at the moment that fresh organic matter is added, contains only microbial tissue. Ignoring the other organic components in the soil may be not as serious as it looks since they are very resistant to microbial attack.
Humus cannot be precisely defined, in general it is assumed that after successive transformations of the material added humus is formed. This is simulated in the following way, the organic matter added partly decomposes forming new microbial tissue, while the decomposition of this microbial tissue is accompanied by a production of humus, which is assumed to remain unaltered.
The decomposition is accompanied by transformations of nitrogen, organic nitrogen from decomposing material is mineralized as ammonium while inorganic nitrogen is incorporated in newly formed microbial tissue. These processes are included in the model by assuming that the inorganic nitrogen necessary, is first obtained from the simultaneously mineralized ammonium and then from other sources already present in the soil. As soon as more inorganic nitrogen is needed than available, the decomposition of nitrogen consuming material is stopped.
Flaig (1967) stated that lignin, after it has been added to the soil, not only decomposes but also absorbs inorganic nitrogen by some kind of chemical process. This process results in a decreasing C/N quotient of the undecomposed fraction, the process is included in the model.

2.2 Description of the simulated processes according to flow charts

The figures 3, 4 and 5 show on flow charts or relational diagrams how the processes are related. The symbols used are explained in Fig. 6. Carbon and nitrogen of organic matter are mineralized into carbon dioxide and ammonia, which may be resynthesized and assimilated by microbes. In decomposition, the pathways of both carbon (Fig. 3 and 4) and nitrogen (Fig. 5) must be considered.
The relative amounts of the four components of the organic matter added to the soil (proteins, sugars, cellulose and lignin) are accounted for separately. In Fig. 3 the amount or level is represented by a rectangle and the abbreviation CMPX of which CMP stands for component and X may be replaced by P for proteins, S for sugars, C for cellulose and L for lignin. The decomposition processes of the four components are similar so that the process presented in Fig. 3 is

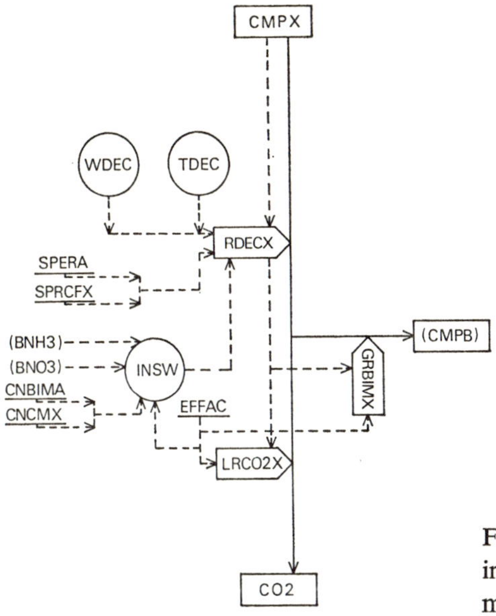

Fig. 3 | Flow chart of carbon in decomposition of organic matter.

applicable (although with different coefficients) to the four components, except that for lignin the absorption of inorganic nitrogen has also to be included (see Section 2.3). Two other amounts or levels of organic material are considered namely biomass (CMPB), representing microbial tissue, and carbon dioxide (CO2). Owing to the decomposition of a component (CMPX) a flow of matter occurs as shown by a solid line. The flows are controlled by their rates, which are symbolized by a valve (see Fig. 6). Thus the decomposition of a component is determined by its rate of decomposition (RDECX), while the simultaneous production of carbon dioxide is controlled by the liberation rate of CO_2 (LRCO2X) and the production of biomass is controlled by the growth rate of biomass (GRBIMX). Several factors influence the actual rates; the convention to represent an information flow by dotted lines is followed here too.
The decomposition of biomass (CMPB) which is represented by solid lines in Fig. 4, forms new biomass (CMPB), humus (HUM) and CO_2 (CO2). The rates controlling the flow of material are rate of decomposition of biomass (RDECB), growth rate of new biomass

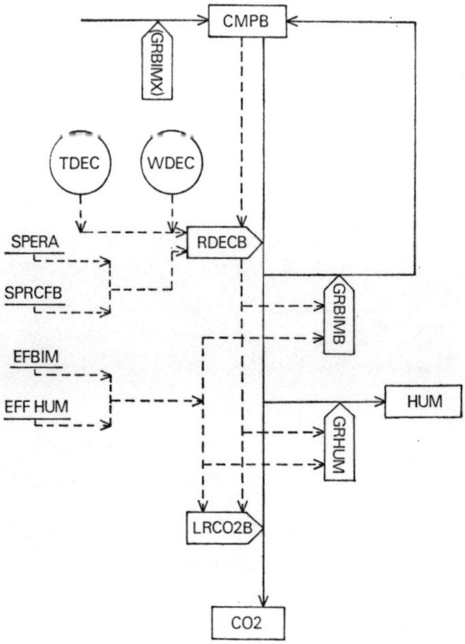

Fig. 4 | Flow chart of carbon in the decomposition of biomass.

(GRBIMB), growth rate of humus (GRHUM) and the production rate of CO_2 (LRCO2B). Notice that in Fig. 4 only the decomposition of biomass is represented while the production of biomass (GRBIMX) accompanying the decomposition of the added components (CMPX) is explained in Fig. 3.

Note that the amounts presented in Fig. 3 and 4 are expressed on a carbon basis, while the quantities in Fig. 5 are expressed on a nitrogen basis. The amount of organic nitrogen incorporated in the added organic matter, biomass and humus is CMPX/CNCMX, CMPB/CNBIMA and HUM/CNHUM respectively. So these quotients are found by dividing the amounts, expressed on a carbon basis (CMPX, CMPB, HUM), by their C/N quotients respectively (CNCMX, CNBIMA, CNHUM).
The inorganic nitrogen can be divided into ammonium (BNH3) and nitrate (BNO3). Decomposition of added organic matter and biomass is accompanied by mineralization of ammonium with rates equal to RDECX/CNCMX and RDECB/CNBIMA respectively. As

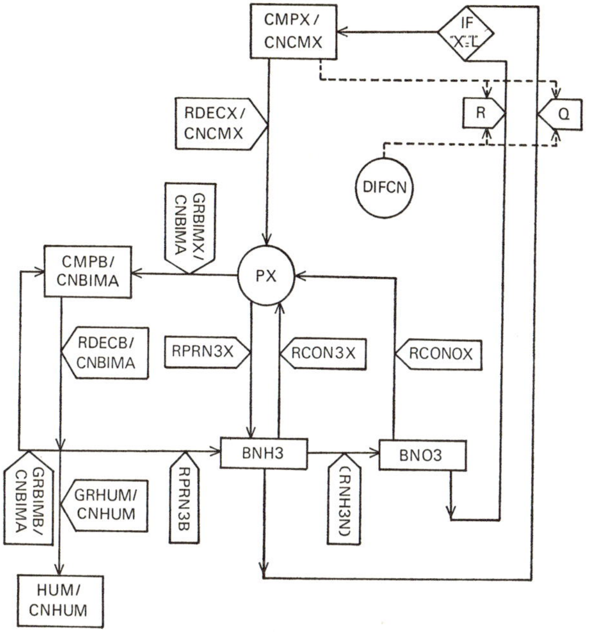

Fig. 5 | Flow chart of nitrogen transformation in soils.

the added organic material decomposes, biomass forms, for which process inorganic nitrogen has to be available. The rate of incorporation of inorganic nitrogen in biomass is represented by GRBIMX/CNBIMA. For the decomposition four possibilities have to be considered:

a RDECX/CNCMX > GRBIMX/CNBIMA

In this situation a surplus of ammonium is produced and the net rate of production of ammonium (RPRN3X) equals the difference of the two rates: (RDECX/CNCMX-GRBIMX/CNBIMA).

b RDECX/CNCMX < GRBIMX/CNBIMA

More mineral nitrogen will be incorporated in the newly formed biomass than simultaneously mineralized. So inorganic nitrogen available from other sources will be used. Harmsen & Kolenbrander

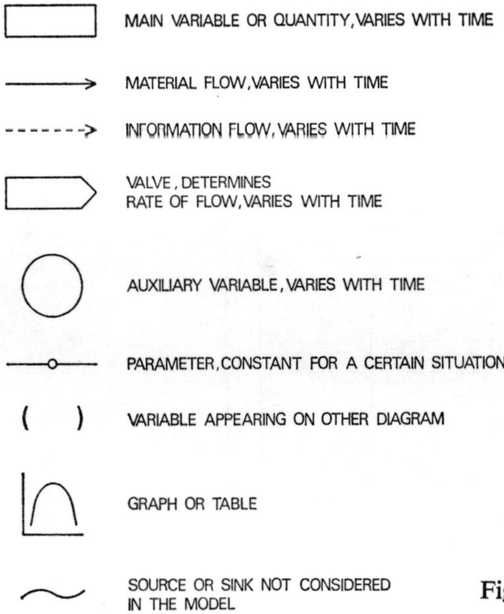

Fig. 6 | Meaning of symbols.

(1965) stated that most investigators agree about the preference of heterotrophic microbes to synthesize their protoplasm from ammonium. Provision has been made to ensure that ammonium is used at first and after that nitrate is consumed. So possibility *b* has been split into:

b1 Ammonium (BNH3) is available in the soil and consequently incorporated in biomass. The rate of consumption of ammonium (RCON3X) equals the difference of the two rates: (RDECX/CNCMX-GRBIMX/CNBIMA).

b2 Nitrate is the only source of inorganic nitrogen available and the rate of consumption of nitrate (RCONOX) defines the process of immobilization of nitrogen. The value of this rate also equals the difference of the two rates mentioned at point b: (RDECX/CNCMX-GRBIMX/CNBIMA).

b3 Inorganic nitrogen is not present, decomposition of organic material (CMPX) stops, so RDECX equals zero.

The ammonium mineralized owing to the decomposition of biomass (rate = RDECB/CNBIMA) is partly incorporated in humus (rate = GRHUM/CNHUM) and partly used for the reproduction of biomass (rate = GRBIMB/CNBIMA). The net rate of production of ammonium owing to the decomposition of biomass (RPRN3B) is accordingly equal to the difference of the rates concerned.
The nitrogen absorption by lignin is shown on the extreme right of Fig. 5. The absorption of ammonium is controlled by the rate Q and the absorption of nitrate by the rate R.

2.3 Definition of the simulated processes in CSMP

In the simulation model six different kinds of organic matter are considered in each soil compartment, namely: humus, proteins, sugars, cellulose, lignin and biomass. All but humus decompose so that it is necessary to define decomposition for each of them. Because the five decomposition processes are similar they can be described by identical equations and accordingly the description will be given in the general form, that is the 'X' form. It is assumed that the rate of decomposition (RDECX, mg C day^{-1} layer^{-1}) is proportional to the amount available (CMPX, mg C layer^{-1}), the proportionality factor is independent of this amount. The proportionality factor is for each of the components (CMPX) equal to the product of the specific rate (SPERA, day^{-1}) and the specific rate coefficient (SPRCFX, dimensionless). The specific rate (SPERA) represents the fraction by which the quantity of proteins (CMPP), the most easily decomposing component, decreases per day owing to decomposition at optimum conditions of the soil layer. The specific rate coefficient (SPRCFX) represents the ratio between the specific rate of each of the four components and that of proteins (SPERA). So the numerical value of the coefficient for proteins (SPRCFP) is equal to unity and for the remaining four components smaller than one. The specific rate coefficient (SPRCFX) is taken to be independent of the water content and temperature of the soil layer, unlike the specific rate (SPERA) which is influenced by both factors. This influence is realized in the program by multiplying the specific rate (SPERA) by WDEC and TDEC representing reduction factors for the decomposition process for water content and temperature respectively. Both factors are expressed as a fraction of the specific rate at optimum conditions. Tabulated functions are used to express the relationship between the temperature of the soil layer (TL, °C) and the temperature reduction factor, TDEC, (dimensionless), and the relationship between the water content reduction factor, WDEC, (dimensionless) and the water holding capacity (WHC, dimensionless). The latter represents the quotient of the actual water content of the soil layer (WCL, dimensionless) and its water content at saturation (WCSAT, dimensionless), expressed as percentage.

As mentioned, two boundary conditions are included, namely: the presence of organic matter and the presence of inorganic nitrogen for the incorporation in biomass if the decomposing substrate does not mineralize enough ammonium to meet the demand. The first boundary condition is defined by the variable SX and the second one by KX. Since it is assumed that five different kinds of organic matter decompose five rates have to be calculated and this is done with a MACRO function (User's manual, IBM, 1967).

```
MACRO RDECX = DEC1 (CMPX,SPERA,SPRCFX,
  SX,KX,WDEC,TDEC,EFX,CNCMX,NIT)              (2.1)
RDEX = CMPX * SPRCFX * SPERA * TDEC *
  WDEC * SX * KX                              (2.2)
KX = INSW ((CNBIMA/EFX - CNCMX),
  FCNSW (NIT,0.,0.,1.),1.)                    (2.3)
SX = FCNSW (CMPX,0.,0.,1.)                    (2.4)
ENDMAC                                        (2.5)
```

The statements 2.1 and 2.5 define the MACRO function while the equations 2.2, 2.3 and 2.4 define the rate of decomposition (RDECX). The CSMP FCNSW function (Eqn 2.4) sets SX to 1 if organic material is present (CMPX > 0), and to zero if no organic material is present (CMPX < 0). If SX is set to zero RDECX becomes zero (decomposition stops) (Eqn 2.2). The CSMP functions INSW and FCNSW set KX (Eqn 2.3) to 1 for two cases, *a* if more ammonium is mineralized than simultaneously incorporated in newly formed biomass (CNBIMA/EFX-CNCMX is greater than 0.) *b* if less ammonium is mineralized than is used for the biomass production (CNBIMA/EFX<CNCMX), but inorganic nitrogen (NIT) is available (FCNSW(NIT,0.,0.,1.)=1). KX is set to zero if no inorganic nitrogen is available. So the values of the variables SX and KX are

equal to one if the boundary conditions are met and equal to zero if the conditions in the soil are not suitable for decomposition.* The variable EFX, the efficiency factor, represents the fraction of decomposed carbon incorporated in the biomass.

To obtain the decomposition rates of the five organic materials for the upper four soil layers the DO-loop form is used:

```
DO 22 I = 1,4
WCL = WC(I)
TL = T(I)
...
...
WHC = WCL/WCSAT * 100.                                  (2.6)
WDEC = AFGEN (WDECTB, WHC)                              (2.7)
FUNCTION WDECTB = (0.,0.),
  (10.,0.02),(30.,0.32),...                             (2.8)
TDEC = AFGEN (TDECTB,TL)                                (2.9)
FUNCTION TDECTB = (0.,0.001),
  (5.,0.15),(10.,0.3),...                               (2.10)
RDECP = DEC1 (CMPP,SPERA,SPRCFP,SP,
  KP,...)                                               (2.11)
RDECS = DEC1 (CMPS,SPERA,SPRCFS,SS,
  KS,...)                                               (2.12)
RDECC = DEC1 (CMPC,SPERA,SPRCFC,SC,
  KC,...)                                               (2.13)
RDECL = DEC1 (CMPL,SPERA,SPRCFL,SL,
  KL,...)                                               (2.14)
RDECB = DEC1 (CMPB,SPERA,SPRCFB,SB,
  KB,...)                                               (2.15)
```

The MACRO function, defined by the statements 2.1–2.5 is used five times (2.11–2.15) for the definition of the rates of decomposition of the proteins (RDECP), sugars (RDECS), cellulose (RDECC), lignin (RDECL), and biomass (RDECB). DO-loop 22 defines the rates for the upper four layers of the soil column (see Fig. 2). In some of the equations three dots are used which means that the equation

* For readers specially familiar with FORTRAN, CSMP provides a sort routine so that KX and SX are calculated first.

considered is presented only partly, a complete description is given in the computer program itself (see Appendix).

A second MACRO function is used for the description of the growth rate of biomass (GRBIMX, mg C day^{-1} $layer^{-1}$), the production rate of CO_2 (LRCO2X, mg C day^{-1} $layer^{-1}$), the rate of production of ammonium (RPRN3X, mg N day^{-1} $layer^{-1}$), the rate of consumption of ammonium (RCON3X, mg N day^{-1} $layer^{-1}$) and the rate of consumption of nitrate (RCONOX, mg N day^{-1} $layer^{-1}$). The growth rate of biomass (GRBIMX) and the production rate of CO_2 (LRCO2X) are proportional to the rate of decomposition (RDECX). The proportionality factors are independent of the decomposition rate. The factors used in the MACRO are the efficiency factor (EFX, dimensionless), which represents the fraction of decomposed carbon incorporated in biomass and the factor (1-EFX) representing the fraction of decomposed carbon used as source of energy by the microbes and consequently oxidized to CO_2 with a gain of energy. For the proteins, sugars, cellulose and lignin the actual factors (EFFAC) and (1-EFFAC) are identical. The corresponding factors for biomass (EFTOT, dimensionless) and (1-EFTOT) differ since a fraction of the decomposed carbon is transformed into humus (see Fig. 4). Hence, the factor EFTOT does not only represent the fraction of decomposed carbon incorporated in newly formed biomass but also the carbon transformed into humus, see Eqn 2.29.
The nitrogen transformations occuring during decomposition of organic material are shown in Fig. 5. The rate of the ammonium production by mineralization is equal to the rate of decomposition (RDECX) divided by the C/N quotient of the decomposing substrate (CNCMX, mg C mg N^{-1}), so RDECX/CNCMX attains the dimension of mg N day^{-1} $layer^{-1}$. The rate of incorporation of mineral nitrogen in biomass is equal to the growth rate of biomass (GRBIMX) divided by the C/N quotient of biomass (CNBIMA, mg C mg N^{-1}), so the quotient GRBIMX/CNCMX has the dimension mg N day^{-1} $layer^{-1}$. The net rate equals the difference of the two previous rates (RDEC/CNCMX-GRBIMX/CNBIMA), but since GRBIMX=RDECX* EFX (Eqn 2.17), the latter difference is rewritten as RDECX* (1/CNCMX-EFX/CNBIMA). To improve readability the variable PX is introduced: PX = (1/CNCMX-EFX/CNBIMA). Only if PX is positive the production of ammonium is possible. The ammonium

production rate (RPRN3X) is calculated with Eqn 2.20 and set to zero if PX is negative. Subsequently if PX is negative, the rate of consumption of ammonium (RCON2X) or of nitrate (RCONOX) is calculated, depending if ammonium (BNH3, mg N layer^{-1}) or only nitrate (BNO3, mg N layer^{-1}) is available. These conditions are introduced by means of INSW and FCNSW functions (Eqns 2.20–2.22).

The general form of the MACRO function is,

```
MACRO GRBIMX,LRCO2X,RPRN3X,RCON3X,
  RCONOX = DEC2(PX,EFX,RDECX,CNCMX,
  CNBIMA,BNH3,BNO3)                          (2.16)
GRBIMX = EFX * RDECX                         (2.17)
LRCO2X = (1. - EFX) * RDECX                  (2.18)
PX = 1./CNCMX - EFX/CNBIMA                   (2.19)
RPRN3X = RDECX * INSW (PX,0.,PX)             (2.20)
RCON3X = RDECX * FCNSW(PX,PX,0.,0.) *
  INSW(BNH3,0.,1.)                           (2.21)
RCONOX = RDECX * FCNSW(BNH3,
  1.,0.,0.) * FCNSW(PX,PX,0.,0.)             (2.22)
ENDMAC                                       (2.23)
```

The INSW function (Eqn 2.20) provides that RDECX is multiplied by PX, if PX is positive. In Eqn 2.21 the INSW and FCNSW functions provide that RDECX is multiplied by PX, if PX is negative, while this product is either multiplied by one if ammonium (BNH3) is present or multiplied by zero if no ammonium is present. The rate defined by Eqn 2.22 equals RDECX multiplied by PX, if at least PX is negative and this product is multiplied by one if no ammonium but only nitrate (BNO3) is present. Note that the presence of nitrate is not mentioned in Eqn 2.22. This is not necessary because if neither ammonium nor nitrate is present, RDECX is set to zero in Eqn 2.2.

The MACRO function DEC2 is used five times to calculate the rates of the five decomposable organic materials. To define the variables for the four layers use is made of DO-loop 22.

```
DO 22 I = 1,4
CMPP = COMP(I)
...
...
GRBIMP,LRCO2P,RPRN3P,RCON3P,RCONOP =
  DEC2(PP,EFFAC,RDECP,...)                    (2.24)
GRBIMS,LRCO2S,RPRN3S,RCON3S,RCONOS =
  DEC2(PS,EFFAC,RDECS,...)                    (2.25)
GRBIMC,LRCO2C,RPRN3C,RCON3C,RCONOC =
  DEC2(PC,EFFAC,RDECC,...)                    (2.26)
GRBIML,LRCO2L,RPRN3L,RCON3L,RCONOL =
  DEC2(PL,EFFAC,RDECL,...)                    (2.27)
GRORMB,LRCO2B,RPRN3B,RCON3B,RCONOB =
  DEC2(PB,EFTOT,RDECB,...)                    (2.28)
```

The use of MACRO function DEC2 implies that the sum of the growth rates of biomass and humus (GRORMB) has been calculated. This latter rate is according to Eqn 2.17 and 2.28 equal to, GRORMB= EFTOT*RDECB. The variable EFTOT represents the sum of the fractions of decomposed carbon incorporated in new biomass (EFBIM, dimensionless) and humus (EFFHUM, dimensionless), starting with biomass itself as substrate. So the growth rate of biomass (REGRBI, mg C day^{-1} layer^{-1}) is obtained with

```
REGRBI = GRORMB - RDECB * EFFHUM              (2.29)
```

Accordingly the growth rate of humus (GRHUM, mg C day^{-1} layer^{-1}) is given by

```
GRHUM = EFFHUM * RDECB                        (2.30)
```

Eventually the net growth rate of biomass per layer (NGRBIL, mg C day^{-1} layer^{-1})

```
NGRBIL = GRBIMP + GRBIMS + GRBIMC +
  GRBIML + REGRBI - RDECB                     (2.31)
```

As stated, lignin absorbs inorganic nitrogen. According to Flaig this process continues till the C/N quotient (CNCML, mg C mg N^{-1}) reaches a value of 20. This process is determined by the rate of ammonium absorption (Q, mg N day^{-1} layer^{-1}) or the rate of nitrate absorption (R, mg N day^{-1} layer^{-1}), the latter one is only calculated if no ammonium (BNH3) is present. It is assumed that the rate of absorption is proportional to the difference of the actual quantity of organic nitrogen in the lignin (CMPL/CNCML, mg N layer^{-1}) minus the quantity which will be reached if the C/N quotient equals 20 (CMPL/20., mg N layer^{-1}). The proportionality factor (actual value −0.0404) is independent of this difference. The rates, defined in CSMP, are obtained with

```
DIFCN = 1./CNCML - 1./20.                                (2.32)
Q = -0.0404 * CMPL *
  FCNSW(DIFCN,DIFCN,0.,0.) *
  FCNSW(BNH3,0.,0., 1.)                                  (2.33)
R = -0.0404 * CMPL *
  FCNSW(DIFCN,DIFCN,0.,0.) *
  FCNSW(BNH3,1.,1.,0.) *
  FCNSW(BNO3,0.,0.,1.)                                   (2.34)
```

If the C/N quotient (CNCML) is greater than twenty, DIFCN is negative (2.32). A FCNSW function (Eqns 2.33 and 2.34) provides then that CMPL is multiplied by DIFCN. If, however, DIFCN is equal to zero or positive, CMPL is multiplied by zero and consequently Q or R are equal to zero. The second FCNSW function (Eqn 2.33) states that absorption is only possible if ammonium (BNH3) is present, while rate R is calculated if no ammonium but nitrate is present.

Hence lignin absorbs inorganic nitrogen with a rate represented by Q or R, whilst it looses organic nitrogen owing to decomposition; the rate of the latter process is equal to RDECL/CNCML. The net rate of nitrogen absorption by lignin (NRNL, mg N day^{-1} layer^{-1}) is obtained by:

```
NRNL = Q + R - RDECL/CNCML                               (2.35)
```

Depending on the net rate (NRNL), the amount of nitrogen present in the lignin (HNL, mg N layer^{-1}) and the C/N quotient (CNCML) change. The amount of nitrogen (HNL) in the lignin is found by integration of the net rate (NRNL). The integration is performed by the CSMP function

```
HNL = INTGRL(IHNL,NRNL)                                  (2.36)
```

The first argument in the integral function (IHNL, mg N layer^{-1}) represents the amount of nitrogen present at the start of the simulation.

The C/N quotient (CNCML) is then obtained with

```
CNCML = CMPL/HNL                                    (2.37)
```

The rate of change of the amounts of proteins (CMPP), sugars (CMPS), lignin (CMPL), cellulose (CMPC) and biomass (CMPB) in the soil layers due to decomposition are calculated by the integral functions:

```
COMP1 = INTGRL(ICMP1,RDC1,4)                        (2.38)
COMP5 = INTGRL(ICMP5,RDC5,4)                        (2.39)
COMP9 = INTGRL(ICMP9,RDC9,4)                        (2.40)
COMP13 = INTGRL(ICMP13,RDC13,4)                     (2.41)
COMP17 = INTGRL(ICMP17,RDC17,4)                     (2.42)
```

The third argument of the integral function indicates that there are 4 integrals, one for each of the upper four soil compartments. The amounts of organic material are stored in arrays so the variable names CMPP, CMPS, CMPL, CMPC and CMPB are replaced by the indexed variables CMP1, CMP5, CMP9, CMP13, and CMP17 in the order given. In the same way the rates of decomposition RDECP, RDECS, RDECL, RDECC and RDECB are replaced by RDC1, RDC5, RDC9, RDC13 and RDC17 respectively. The first argument in the integral function represents the initial amounts of a component in the upper four soil compartments, also stored in arrays. These initial amounts (ICMP, mg C $layer^{-1}$) in each of the four compartments are obtained with

```
DO 4 I = 1,4
ICMP(I) = PROT * VOLL(I)                            (2.43)
ICMP(I+4) = SUGAR * VOLL(I)                         (2.44)
ICMP(I+8) = LIGN * VOLL(I)                          (2.45)
ICMP(I+12) = CELLU * VOLL(I)                        (2.46)
ICMP(I+16) = BIMA * VOLL(I)                         (2.47)
```

The variables PROT, SUGAR, LIGN, CELLU, and BIMA represent the amounts of proteins, sugars, cellulose, lignin and biomass added or present in a soil at the start of the simulation. Their concentration is expressed as carbon equivalent in mg cm^{-3} and multiplied by the volume of the layer (VOLL, cm^3) to obtain the mass of carbon equivalent in each layer (mg C $layer^{-1}$) (ICMP).

2.4 Approximations for input variables

Van Schreven (1964) recorded data about the quantity of carbon dioxide evolved and the change in the level of inorganic nitrogen in the soil during decomposition of wheat straw at known conditions of moisture and temperature. Wheat straw was added to the soil at a rate of 28500 kg per hectare which corresponds to 4670 kg C per hectare. In accordance with general information about wheat straw, this amount was divided in 4 fractions, proteins, sugars, cellulose and lignin. By taking into account the nitrogen content of wheat straw and a C/N quotient for protein of 6, it was calculated that the protein dressing was 0.5 mg C cm^{-3} soil, so PROT = 0.5 mg/cm^3. For the other components these values are, SUGAR = 1.0, CELLU = 1.8, LIGN = 1.3.
The specific rate of decomposition (SPERA) is related to the decomposition of proteins. It is expressed as the fraction of the amount decomposed per day, at optimum conditions of moisture and temperature. Since each component has its own specific rate of decomposition a so called specific rate coefficient factor for each component (SPRCFX) is introduced. If this variable is set to 1 for proteins (SPRCFP), the other coefficients are: 0.75 for sugars (SPRCFS), 0.05 for cellulose (SPRCFC) and 0.0095 for lignin (SPRCFL). So the product of SPERA and SPRCFX represents the fraction of the component which decomposes per day. The coefficients for proteins, sugars and cellulose were evaluated from data of van Schreven (1964) by trial and error. Data of Mayaudon & Batistic (1970) were used to calculate the coefficient for lignin (SPRCFL).
The value of the same coefficient for biomass (SPRCFB) was taken as 0.5, while the initial amount of biomass (BIMA) was estimated at 0.04 mg C cm^{-3} with a C/N quotient of 8 (CNBIMA = 8). The values for the C/N quotients of proteins, sugars, cellulose and lignin are CNCMP = 6, CNCMS = 200, CNMC = 200 and CNCML = 100 respectively.

The efficiency factor (EFFAC) represents the fraction of decomposed carbon incorporated in newly formed biomass and the factor (1-EFFAC) represents the carbon oxidized to CO_2. The value of EFFAC was calculated from the following data of van Schreven

(1964): During 1 week of incubation of a soil, mixed with wheat straw, 286 mg of carbon, originating from the straw, evolved as CO_2. After 1 week 7.5 mg of nitrogen had been immobilized. Suppose that during this week of incubation X mg carbon of the straw is incorporated in newly formed biomass with a C/N quotient of 7. So $X/7$ mg nitrogen is used, while simultaneously $(286+X)/45.9$ mg N is mineralized owing to decomposition of the wheat straw (C/N quotient straw = 45.9). Since 7.5 mg N had been immobilized the total amount of nitrogen used for biomass production equals $(286+X)/45.9+7.5$ and $X/7 = (286+X)/45.9+7.5$. So $X = 113$ mg C. Thus in 1 week 399 mg C (286+113) has been decomposed of which 113 mg C is used for biomass. This gives an efficiency factor (EFFAC) of 113/399 = 0.28. Because of a lack of information it is assumed that the efficiency factor is the same for proteins, sugars, cellulose and lignin.

In the model the decomposition of biomass causes the production of humus (HUM), with a C/N quotient of 10 (CNHUM = 10), CO_2 (CO2) and biomass (CMPB), see also Fig. 4. The efficiency factor for humus (EFFHUM) and the efficiency factor for biomass (EFBIM) represent the fraction of carbon incorporated in humus and newly formed biomass respectively. The values of the factors were estimated from data of Mayaudon & Simonart (1963). In their experiment an amount of Azotobacter had been added to the soil of which 40% remained in the soil after 60 days. So 60% had been mineralized to CO_2. Since the quotient immobilized nitrogen/mineralized nitrogen equalled 0.4 in their experiment a fraction of 24% (0.4 × 60%) of the material was humified and 16% (100–60–24) remained as living matter. According to this EFFHUM = 0.24 and EFBIM = 0.16. In the model the humus (HUM) produced is thought to be resistent to any further microbial attack, thus it stays unaltered in the soil. The amount of humus present in the soil at the start of the simulation was set to zero (Eqn 21), so that no decision had to be made about which fraction of the organic material already present in the soil consisted of humus.

The specific rate of decomposition (SPERA = 0.1) is defined for optimum conditions. The actual rate also depends on moisture and temperature. Since, however, both moisture and temperature fluctuate a provision was made to reduce the specific rate (SPERA) as a

Table 1 Factors to account for the influence of the water content and temperature on microbial processes

Relative water content (WHC) (in % of the content at saturation)	Value of WDEC	Soil temperature (TL) (degrees Centigrade)	Value of TDEC
0	0	0	0.001
10	0.02	5	0.15
30	0.32	10	0.30
40	0.65	20	0.64
50	0.81	30	0.81
60	1.00	40	0.96
70	0.80	50	1.00
80	0.55		
90	0.48		
100	0.45		

function of WDEC and TDEC. They represent the reduction factor for water content and temperature respectively. The two reduction factors have values varying between zero and one. Data of Kononova (1961) were used to calculate the reduction factors (Table 1).
In the preceding section it is stated that lignin decreases its C/N quotient by absorbing inorganic nitrogen. The rate Q (Eqn 2.33) controls the absorption of ammonium, while R (Eqn 2.34) controls the absorption of nitrate. Both rates may be represented by the following differential equation:

$$Q = R = d(HNL)/dt = k \times (1/CNCML-1/20) \times CMPL$$

The variable HNL represents the amount of nitrogen incorporated in the lignin while CMPL represents the amount of lignin expressed on a carbon basis and CNCML the C/N quotient. The value of the proportionality factor k has been evaluated with

$$d(HNL)/dt = k \times (CMPL/CNCML-CMPL/20)$$

since CMPL/CNCML=HNL

$$d(HNL)/dt = k \times (HNL-CMPL/20)$$

This equation can be rewritten as

$$d(HNL)/(HNL-CMPL/20) = k \times dt$$

Since lignin decomposes slowly, it is assumed that the quantity (CMPL) may be considered to be constant for a certain time*. Now the differential equation can be integrated:

$$\ln(HNL-CMPL/20) = k \times t + \text{constant}$$

This latter equation can be replaced by $(HNL-CMPL/20) = e^{kt} \times K'$, in which K' is an integration constant. At time $(t) = 0$, $HNL = CMPL/CNCML = 1250/100 = 12.5$; substitution of this value in $(HNL-CMPL/20) = e^{kt} \times K'$ shows that K' equals -50.
At time $(t) = 70$, $CMPL = 1250$ and $CNCML = 21$. Substitution results in the value of the proportionality constant k:

$$(1250/21 - 1250/20) = -50 \times e^{k \times 70}; \text{ and } k = -0.0404.$$

* CMPL is only taken constant for the calculation of the proportionality factor k. In the CSMP program it is of course a variable.

3 Nitrification process

3.1 Basic process

Nitrate as an anion is hardly adsorbed by the soil complex and so is intensively leached; ammonium can be adsorbed by the negatively charged soil complex and thus is leached more slowly. Therefore nitrate (BNO3) and ammonium (BNH3) are considered separately in the simulation model. The real situation is further simplified by assuming that nitrate moves along with water in the soil and that ammonium is adsorbed completely. Thus no exchange reaction between NH_4^+ adsorbed and NH_4^+ in solution is taken into account. However a 4-ion exchange program, written in the same way has been worked out by Reiniger et al (1972), so that exchange could be introduced if required. From their work it can otherwise be concluded that the movement of NH_4^+ compared to the movement of NO_3^- can indeed be neglected. Leaching of nitrate can cause a considerable loss of nitrogen, so it is necessary to include the processes transforming ammonium to nitrate. In soil, ammonium may be nitrified to nitrate by a few species of soil microbes, that are the autotrophic organisms, the bacteria Nitrosomonas, Nitrosococcus and Nitrobacter. Nitrosomonas and Nitrosococcus use ammonium as energy source by oxidizing it to nitrite while Nitrobacter oxidize in their turn nitrite to nitrate, when oxygen is available. It is generally accepted that under normal soil conditions the rate of oxidation of nitrite to nitrate is much faster than the oxidation of ammonium to nitrite and thus the latter process is the rate limiting step in the nitrification process. Consequently, in this simple model, ammonium is nitrified directly to nitrate, while the rate is determined by the conversion of ammonium into nitrite. The above-mentioned microbes also absorb ammonium for the synthesis of their own tissue. However, this amount is negligible compared with the quantity used as a source of energy.

In a resting soil there will be an initial population of microbes contributing to nitrification. They start growing either immediately after the addition of ammonium or there is a certain delay. In this model

the so called ammonium oxidizers are assumed to grow as soon as ammonium is available, mainly because data on the delay are so complex that a general trend could not be recognized. The growth is assumed to be exponential. The death rate for the microbes is taken as proportional to the number of micro-organisms, i.e. death is supposed to occur at random.
Perfusion experiments of Quastel & Scholefield (1951) showed a maximum rate of nitrification. This seems to indicate that the population does not increase indefinitely but rises to a maximum value, which is thought to be determined by the total particulate surface available. Therefore provision has been made in the model that the rate of nitrification does not exceed a maximum value.
Many soil characteristics influence nitrification, aeration status, pH, temperature and water content of the soil. Variations in the aeration status or pH are neglected but the influence of temperature and mosture is considered.

3.2 Description of the simulated process according to the flow chart

The qualitative relationships of the nitrification process are shown in Fig. 7. The amount of ammonium (BNH3) in the soil may decrease by a flow of ammonium, caused by nitrification: This flow from BNH3 (ammonium) towards BNO3 (nitrate) is described by the rate of nitrification of ammonium (RNH3N). As shown by the broken lines the rate (RNH3N) is influenced by such factors as the temperature coefficient (TCOF) and water content coefficient (WCCOF). These coefficients reduce the rate (RNH3N) as a function of the actual temperature and water content. The number of microbes which contribute to nitrification, called ammonium oxidizers (AMOX), with their specific rate of oxidation (RAMOX), also influence the rate (RNH3N). A maximum rate of nitrification of ammonium (MRNH3N) is introduced so that the nitrification rate (RNH3N) does not exceed a maximum value.
The number of ammonium oxidizers (AMOX) may increase by growth of the species, controlled by the growth rate of ammonium oxidizers (GRAMOX). GRAMOX depends on the nitrification rate RNH3N and so indirectly on the number of ammonium oxidizers (AMOX). GRAMOX is defined by the quotient (RNH3N/RAMOX), and the growth rate constant of ammonium oxidizers (GRCAO).

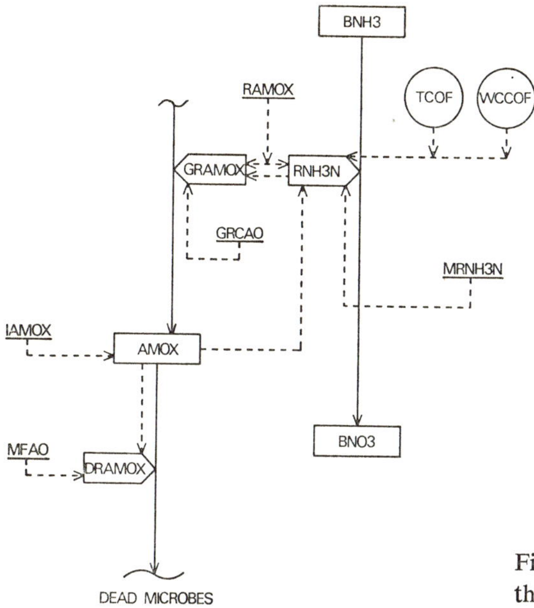

Fig. 7 | Flow chart of the nitrification process.

A decrease in the number of ammonium oxidizers (AMOX) is controlled by the death rate of ammonium oxidizers (DRAMOX). DRAMOX depends on the number of microbes (AMOX) and the mortality factor of the ammonium oxidizers (MFAO).

3.3 Definition of the simulated process in CSMP

The rate of nitrification of ammonium (RNH3N, mg N day^{-1} $layer^{-1}$) is taken as proportional to the number of ammonium oxidizers (AMOXL, number $layer^{-1}$). The proportionality factor is independent of the amount of substrate, i.e. ammonium, and is represented by the variable RAMOX (mg N day^{-1} $number^{-1}$) (Quastel & Scholefield 1951). RAMOX expresses the rate of nitrification of a single ammonium oxidizer at optimum conditions of temperature and water content. An evaluation of its value from experimental data was complicated (see Section 3.4). The specific rate (RAMOX) is multiplied by the temperature coefficient (TCOF, dimensionless) and water content coefficient (WCCOF, dimensionless) to provide a reduction if the conditions are not optimum. The multiplication of TCOF and WCCOF implies that the influence of the temperature on nitrification is independent of the actual water content. Both factors have a value varying between zero and one and are expressed as a fraction of the rate at optimum conditions. Tabulated functions are used to express the relationship between TCOF and temperature (TL, °C) and WDEC and the water holding capacity of the soil layer (WHC, dimensionles). The following equations are used for the upper 4 layers:

```
DO 22 I = 1,4
WCCOF = AFGEN(WCNFTB,WHC)                                  (3.1)
FUNCTION WCNFTB = (0.,0.05),
  (10.,0.1),(20.,0.17)...                                  (3.2)
TCOF = AFGEN(TPNFTB,TL)                                    (3.3)
FUNCTION TPNFTB = (0.,0.),(10.,0.12),
  (20.,0.17)...                                            (3.4)
RNH3N = AMIN1(AMOXL * RAMOX *
  INSW(BNH3,0.,1.),MRNH3L) * TCOF *
  WCCOF                                                    (3.5)
```

The INSW function (Eqn 3.5) states that RAMOX is either multiplied by 1 if ammonium (BNH3) is available, or multiplied by zero if ammonium is not available. The AMIN1 function (Eqn 3.5) provides that the smallest value of its arguments, i.e. either AMOXL * RAMOX * INSW(BNH3,0.,1.) or the maximum rate of nitrification per layer (MRNH3L, mg N day^{-1} $layer^{-1}$) is multiplied by TCOF * WCCOF.

The growth rate of ammonium oxidizers (GRAMOX, number day^{-1} $layer^{-1}$) is obtained with:

```
GRAMOX = GRCAO * RNH3N/RAMOX                               (3.6)
```

This definition of the growth rate (Eqn 3.6) provides that since the nitrification rate (RNH3N) is influenced by the temperature and water content of the soil, the growth rate of the bacteria is also influenced by these factors. The growth rate constant (GRCAO, day^{-1}) is a proportionality factor.

The death rate of ammonium oxidizers (DRAMOX, number day^{-1} $layer^{-1}$) equals:

```
DRAMOX = AMOXL/MFAO                                        (3.7)
```

MFAO (day) is a mortality factor. The quotient 1/MFAO represents the fraction of microbes which die per day. The use of the reciprocal value of 1/MFAO is chosen because this is common practice among bacteriologists.
The net growth rate (NGRAMX, number day^{-1} layer^{-1}) is obtained with

```
NGRAMX = GRAMOX - DRAMOX
```
(3.8)

The number of ammonium oxidizers in each of the upper four soil layers is obtained with

```
AMOX1 = INTGRL(IAMOX1,NGRAM1,4)
```
(3.9)

The net growth rates (NGRAMX) of the upper four soil layers are stored in an array defined by NGRAM, just like the initial number of ammonium oxidizers (IAMOX, number layer^{-1}) are stored in the array IAMOX.
So far all the rates of the transformation processes of nitrogen have been defined (Fig. 5; Fig. 7). So the production rate of ammonium per layer (PRN3L, mg N day^{-1} layer^{-1}), the consumption rate of ammonium per layer (CRN3L, mg N day^{-1} layer^{-1}), the consumption rate of nitrate per layer (CRNOL, mg N day^{-1} layer^{-1}) and the net rate of formation of ammonium per layer (NFNH3L, mg N day^{-1} layer^{-1}) are defined by

```
PRN3L = RPRN3P + RPRN3S + RPRN3C +
  RPRN3L + RPRN3B                         (3.10)
CRN3L = RCON3P + RCON3S + RCON3C +
  RCON3L + RCON3B - Q                     (3.11)
CRNOL = RCONOP + RCONOS + RCONOC +
  RCONOL + RCONOB - R                     (3.12)
NFNH3L = PRN3L + CRN3L - RNH3N            (3.13)
```

The amount of ammonium (BNH3, mg N layer^{-1}) for each of the four layers is found by integration of the rates with

```
NH3B1 = INTGRL(INH3B1,NFNH31,4)
```
(3.14)

The variable names NH3B1, INH3B1, and NFNH31 represent the names of the arrays containing the amounts of ammonium (BNH3), initial amounts of ammonium (INH3B, mg N per layer) and the net rates (NFNH3L) of the upper four soil layers.

3.4 Approximations for input variables

The nitrification process as described has to be considered as a rather simplified version of the process occuring in nature. Despite these simplifications it has been difficult to find suitable data in literature for the approximation of the variables which define the specific growth rate constant of ammonium oxidizers (GRCAO), the specific rate of nitrification of ammonium oxidizers (RAMOX), the mortality factor of ammonium oxidizers (MFAO) and the initial number of ammonium oxidizers (IAMOX).
The variable (GRCAO) is used for the definition of the growth rates of ammonium oxidizers (GRAMOX), as represented by Eqn 3.6. GRAMOXL is derived from differential equation, d(AMOXL)/dt= GRCAO × AMOXL, in which t represents time and RNH3N/RAMOX (Eqn 3.6) is replaced by AMOXL, although this is only permissible at optimum conditions of temperature and water content. After integration and evaluation of the integration constant the following equation is obtained:

$$\mathrm{AMOXL} = \mathrm{IAMOX} \times e^{\mathrm{GRCAO} \times t}$$

IAMOX represents the number of ammonium oxidizers at $t = 0$. In bacteriology it is common practice to calculate the value of GRCAO for AMOXL = 2 × IAMOX. Thus if t represents the generation time, GRCAO = ln (2)/(generation time). Stojanovic & Alexander (1958) found a generation time of 32 h for Nitrosomonas bacteria. This corresponds with a value of 0.022 h^{-1} or about 0.5 day^{-1} for GRCAO. The values of the specific rate of nitrification (RAMOX) and the maximum rate of nitrification of ammonium (MRNH3N) have also been estimated from data of Stojanovic & Alexander. They measured the amount of ammonium nitrified due to percolation of a solution of ammonium sulphate through a soil column. When a solution with an NH_4-N concentration of 1800 μg/ml was percolated for 20 days at 30 °C, the rate of nitrification increased exponentially during the first 10 days and remained almost constant for the next 10 days (Stojanovic & Alexander, Fig. 2; 1958). In the model a constant nitrification rate represents the situation in which the soil is saturated with ammonium oxidizers. So the constant rate of nitrification derived from their data represents the maximum rate of nitrification of ammonium (MRNH3N) for that particular soil. It amounts to about 70 μg N/g

soil, which is equivalent to $70 \times 1.3/1000$ mg N/cm^3 soil, when the soil bulk density = 1.3. According to Alexander (1961), an enriched soil contains about 2.1×10^6 Nitrosomonas bacteria per g soil and if it is assumed that the percolated soil is enriched, the specific rate of nitrification (RAMOX) equals about 3.3×10^{-5} μg N nitrified per day per ammonium oxidizer ($70/2.1 \times 10^6$). It is stressed again that the values of the variables are only approximations.
The death rate of ammonium oxidizers (DRAMOX) has been derived from the differential equations:

$$d(AMOXL)/dt = -AMOXL/MFAO$$

After integration and evaluation of the integration constant the equation becomes

$$AMOXL = IAMOX \times e^{-t/MFAO},$$

in which IAMOX = number of ammonium oxidizer at time $t = 0$ and MFAO = mortality factor (days). This equation has been rewritten to solve MFAO:

$$MFAO = t/\ln(IAMOX/AMOXL).$$

In the model the value of MFAO represents the time taken for a number of microbes to decay to a fraction equal to 1/e of the original value, so

$$\ln(IAMOX/AMOXL) = 1.$$

It follows that the value of MFAO can be determined from decay curves of Nitrosomonas bacteria. Unfortunately no suitable data have been found in literature so far. Hence the value of MFAO = 30 days has to be considered as an arbitrary value.
Data of Russel et al (1925, Table 2) have been used to describe the influence of temperature on nitrification. The temperature coefficient (TCOF) has been expressed as a fraction of the rate of nitrification at optimum temperature.
Data of Greaves & Carter (1920) have been used for the influence of the water content on the nitrification rate. Also, the water content coefficient (WCCOF) has been expressed as a fraction of the nitrification rate at optimum water content (Table 3).
Several authors have submitted data for the influence of the temperature and water content of the soil on the nitrification process. Accord-

Table 2 Factors to account for the influence of the temperature on the rate of nitrification

Temperature (degrees Centigrade)	Production of nitrate p.p.m.	TCOF
6.9	2.7	0.12
15.1	3.7	0.16
22.9	8.1	0.36
25.4	8.7	0.39
34.5	22.5	1.00
48.3	14.8	0.66
55.0	2.2	0.10

Table 3 Factors to account for the influence of the water content on the rate of nitrification.

Relative water content (WHC) (in % of the content at saturation)	WCCOF
10	0.11
30	0.31
50	0.86
60	1.00
70	0.40
80	0.10
100	0.0

ing to Kolenbrander (1969) the difference between the results was rather small, so that the choice of data was arbitrary.

4 Transport processes

4.1 General

The transport processes included in the model are transport of heat, water and nitrate. It is assumed that the ammonia is totally adsorbed by the soil complex as NH_4^+. In their book de Wit & van Keulen (1972) have explained the principle of transport processes and the theoretical considerations involved, and have described the program in CSMP fully. Thus only a brief description of the transport processes will be given here.

4.2 Flow of heat

In the model the flow of heat may occur by diffusion and/or by the transport of water with another temperature as its surroundings. It is calculated for the upper ten layers (see Fig. 1) of the soil. For the layers 11–20 de Wit & van Keulen showed changes in temperature, caused by the transport of heat and water, which are so small that they can be ignored. Thus the temperatures of these layers remain unchanged.
The temperature of the layers 1–10 is found by dividing the volumetric heat content by the volumetric heat capacity per layer.
At the start of the simulation run (time = 0) the initial volumetric heat capacity (IHECA, cal cm^{-3} °C) is obtained from the heat capacity of minerals (HECAM, cal cm^{-3} °C), the relative amount of minerals per cm^3 soil (AMMIN), the initial water content (IWC, cm^3 cm^{-3}) and the heat capacity of water (HECAW, cal cm^{-3} °C).

```
IHECA = AMMIN * HECAM + IWC * HECAW        (4.1)
```

The initial heat content (IHCL, cal) for the upper ten layers is obtained from the initial temperature (INTEMP, °C) of the upper ten layers, the volume of the layers (VOLL, cm^3) and IHECA.

The calculation is realized in DO-loop 3

```
DO 3 I = 1,10
IHCL(I) = IHECA * VOLL(I) * INTEMP       (4.2)
3 CONTINUE
```

The values of these variables are, for the situation which is described in the Appendix, INTEMP = 10, IWC = 0.15, HECAM = 0.48, HECAW = 1, AMMIN = 0.57. The following variables are needed to update the system at every time step. Heat flow caused by the flow of water (HEFLW, cal day^{-1} cm^{-2}), depends on the flow rate of water (FLRW, cm day^{-1}) crossing the boundary of two adjacent layers, the temperature of the flowing water (T, °C) and the heat capacity of the water (HECAW). The flow into the first layer is given by

```
HEFLW(1) = FLRW(1) * HECAW * TW       (4.3)
```

The water entering the first layer originates from irrigation water with a temperature represented by TW (°C), in this example TW is taken as 16 °C. The heat flow into the next nine layers is calculated in DO-loop 17:

```
DO 17 I = 2,10
HEFLW(I) = FLRW(I) * HECAW *
  INSW(FLRW(I),T(I),T(I-1))       (4.4)
17 CONTINUE
```

INSW is a CSMP statement which defines that HECAW is multiplied by T(I-1) if the water flows downwards and by T(I) if the water flows upwards. The heat flow out of the tenth layer is given by

```
HEFLW(11) = FLRW(11) * HECAW * T(10)       (4.5)
```

The flow of heat by diffusion (HEDI, cal day^{-1} cm^{-2}) depends in general on the temperature gradient and the heat conductivity (HECO, cal cm^{-1} day^{-1} °C^{-1}). HECO depends on the water content (WC). The relationship between HECO and WC is given by FUCNCTION HECOF. The value of HECO as a function of the water content of the layers 1–11 is found by interpolation in DO-loop 18:

```
DO 18 I = 1,11
HECO(I) = AFGEN(HECOF,WC(I)) *
  24. * 3600./1000.                           (4.6)
18 CONTINUE
FUNCTION HECOF = (0.,0.5),
  (0.05,0.6) ...
```

The factor (24 * 3600/1000) is necessary to provide the right dimension for HECO, since values of the heat conductivity in the FUNCTION HECOF are expressed as mcal cm^{-1} sec^{-1} $°C^{-1}$.

Data of van Duin (1956, Fig. 11) have been used to express the influence of the water content on the heat conductivity for a soil (clay) with a volume fraction of solid material equal to 0.55.

The temperature gradient is defined as (T(I-1)-T(I))/DIFD(I), in which DIFD represents the distance (cm) between the centres of two adjacent layers. Since the gradient is calculated from the difference in temperature between two layers the appropriate value of the average heat conductivity (AVHECO, cal cm^{-1} day^{-1} $°C^{-1}$) for each of the layers 2–11 is found from

```
DO 19 I = 2,11
AVHECO(I) = (TCOM(I-1) +
  TCOM(I))/(TCOM(I-1)/HECO(I-1) +
  TCOM(I)/HECO(I))                            (4.7)
19 CONTINUE
```

TCOM is the thickness of a compartment (4 cm), note that the average heat conductivity is calculated in the same way as an average electric conductivity.

The heat flow by diffusion (HEDI) for the layers 2–10 is given by:

```
DO 20 I = 2,10
HEDI(I) = AVHECO(I) * (T(I-1)-
  T(I))/DIFD(I)                               (4.8)
20 CONTINUE
HEDI(11) = 0.                                 (4.9)
```

The flow of heat into the first layer (HEDI(1)) depends on the air temperature (AIRT, °C) and the heat conductivity of the first layer (HECO(1)).

The air temperature is given by a forcing function, i.e. a function which states how the temperature changes with time, independent of the temperature of the underlying soil. In the FUNCTION AIRTF the daily variation of the air temperature is given. It repeats itself daily (i.e. each time that TIME is a whole number) by means of an AMOD function.

```
B = AMOD(TIME,1.)                                   (4.10)
AIRT = AFGEN(AIRTF,B)                               (4.11)
FUNCTION AIRTF = 0., 10.,0.25,
  12.,0.50,20.,0.58,22.,0.75,18.,
  1.,10.                                            (4.12)
HEDI(1) = HECO(1) * (AIRT - T (1))/
  DIFD(1)                                           (4.13)
```

The net heat flow (NHFL, cal day^{-1} cm^{-2}) of a layer is given by the difference of inflow and outflow of heat for that layer.

```
DO 26 I = 1,10
NHFL(I) = HEFLW(I)-HEFLW(I+1)+
  HEDI(I)-HEDI(I+1)                                 (4.14)
26 CONTINUE
```

The heat content of a layer (HCL, cal) for layers 1–10 is obtained with

```
HCL1 = INTGRL(IHCL1,NHFL1,10)                       (4.15)
```

and the temperature of the layer is obtained by dividing the heat content (HCL) by the heat capacity of the layer.

```
DO 9 I = 1,10
T(I) = HCL(I)/((AMMIN* HECAM+ WC(I)*
  HECAW)*VOLL(I))                                   (4.16)
9 CONTINUE
```

4.3 Flow of water

The law of conservation of matter and Darcy's law for the flow of water provide the following differential equation that defines water flow in the soil profile in a vertical direction:

$v = D\,d\theta/dx + K$,

in which v represents the flow rate of water (cm day^{-1}), D the diffusivity (cm^2 day^{-1}), θ the water content (cm^3 cm^{-3}), x distance in cm (downward direction positive value), and K the conductivity (cm day^{-1}).

De Wit & van Keulen (1972) wrote the above differential equation as a difference equation for each layer of the soil profile.
For the layers 2–20 the following equations are obtained:

```
DO 10 I = 2,20
AVDIF(I) = (W * D(I-1) +
  D(I))/(W + 1.)                                  (4.17)
AVCOND(I) = (W * COND(I-1) +
  COND(I))/(W+1)                                  (4.18)
FLRW(I) = AVDIF(I) * (WC(I-1)-
  WC(I))/DIFD(I)+AVCOND(I)                        (4.19)
10 CONTINUE
```

Since the flow rate of water (FLRW, cm day^{-1}) is calculated in the soil at the boundary of two adjacent layers an average diffusivity (AVDIF, cm^2 day^{-1}) is calculated from the diffusivities (D, cm^2 day^{-1}) of the two particular layers (Eqn 4.17).
A weighted average can be calculated by means of variable W. However in the model this procedure has been avoided since W is set to 1. Equation 4.18 defines the average conductivity (AVCOND, cm day^{-1}) from the conductivities (COND, cm day^{-1}) of the two adjacent layers. Because both the diffusivity and conductivity depend on the water content (WC, cm^3 cm^{-3}) FUNCTIONS are included which give the relationships of water content to the diffusivity and to conductivity.

```
DO 8 I = 1,20
D(I) = AFGEN(DTBL,WC(I))                          (4.20)
COND(I) = AFGEN(CONDTB,WC(I))                     (4.21)
8 CONTINUE
FUNCTION DTBL = (0.02,4.),(0.05,1)...             (4.22)
FUNCTION CONDTB = (0.018,1.E-8),
  (0.03,1.E-7)...                                 (4.23)
```

In the CSMP program (see Appendix) the relationships of the water content (WC) to the diffusivity (D) and to conductivity (COND) have been provided up to the saturated water content of the soil. At saturation the water content (WC, cm^3 cm^{-3}) equals 0.43. Data of Staple (1969) have been used for the relationship between water content (WC) of the soil (Castor loam) and diffusivity (D) and conductivity (COND).
So far the flow rates have been defined for the layers 2–20

(FLRW(2)-FLRW(20)). Owing to different boundary conditions the flow rates FLRW(1) and FLRW(21), are calculated separately, see also Fig. 2.
Since water is supplied to the surface by irrigation it depends on the irrigation scheme (Section 4.3.1) if water is available or not.
If water is available at the surface, FLRW(1) follows from:

```
AVDIF(1) = (W * DSRF + D(1))/
  (1. + W)                                   (4.24)
AVCOND(1) = (W * CONDSF + COND(1))/
  (1. + W)                                   (4.25)
FLRW(1) = AVDIF(1) * (WCSAT - WC(1))/
  DIFD(1) + AVCOND(1)                        (4.26)
```

Both the diffusivity of the surface (DSRF) and the conductivity of the surface (CONDSF) are considered to be determined at saturated water content (WCSAT).
The value of DSRF is derived from FUNCTION DTBL (Eqn 4.22) and that of CONDSF from FUNCTION CONDTB (Eqn 4.23) at a value of WCSAT = 0.43.
If no water is available at the surface:

```
FLRW(1) = 0.
```

For the definition of the flow rate of water over the last boundary (FLRW(21)) it is assumed that the water content of the soil profile deeper than layer 20 (WCLAST) remains unaltered. Also the diffusivity (DSL) and conductivity (CONDL) for this last layer are derived from the FUNCTIONS DTBL (Eqn 4.22) and CONDTB (Eqn 4.23), respectively. In the model WCLAST equals 0.15.

```
AVDIF(21) = (W * D(20) + DSL)/
  (W + 1.)                                   (4.27)
AVCOND(21) = (W * COND(20) + CONDL)/
  (W + 1.)                                   (4.28)
FLRW(21) = AVDIF(21) * (WC(20) -
  WCLAST)/(0.5 * TCOM(20)) +
  AVCOND(21)                                 (4.29)
```

According to Eqn 4.29 FLRW(21) may be positive or negative depending on the difference (WC(20)-WCLAST), however, a restriction is made so that FLRW(21) can be positive only. This restriction is obtained by an INSW function:

```
FLRW(21) = INSW(FLRW(21),0.,
  FLRW(21))                                    (4.30)
```

The net flow rate of water (NFLW, cm day^{-1}) for any layer equals the difference between the inflow rate and outflow rate.

```
DO 25 I = 1,20
NFLW(I) = FLRW(I) - FLRW(I+1)                  (4.31)
25 CONTINUE
```

The volume of water (VOLW, cm^3) for each layer is obtained with

```
VOLW1 = INTGRL(IVLW1,NFLW1,20)                 (4.32)
```

in which IVLW represents the initial volume of water (cm^3).
The water content (WC, cm^3 cm^{-3}) follows from

```
WC(I) = VOLW(I)/VOLL(I)                        (4.33)
```

The total inflow (INFLWW, cm^3) and total outflow (OUTFLW, cm^3) is calculated with

```
INFLWW = INTGRL(0., FLRW(1))                   (4.34)
OUTFLW = INTGRL(0., FLRW(21))                  (4.35)
```

4.3.1 *Irrigation scheme*

Only a brief description of the irrigation scheme has been included here. For the equations see the CSMP program in the Appendix. The meaning of the variables is explained by comments in the program. With the irrigation scheme it is possible to irrigate at regular time intervals. The interval selected here is 14 days, described by

```
QQ = AMOD(TIME, PIRR)
```

in which the irrigation frequency PIRR = 14 days. Irrigation takes place at ON (= day 5.33) over DIRR (= 0.31) days.
The irrigation may also be stopped after a certain time. In the programme provision is made that in fact only one irrigation cycle takes place. The maximum amount of water that can be supplied (WAIR, cm) with an intensity (INTENS) of 1 cm h^{-1} is equivalent to 7.5 cm. Subsequent to the supply of irrigation water the transport

rate to layer (FLRW(1)) is calculated with eqns 4.24–26. If FLRW(1) is smaller than the intensity of the irrigation (INTENS), ponding will occur. Provision has been made for the irrigation to stop if the thickness of the water layer formed exceeds the value of the maximum permitted hydraulic head (HMAX = 0.3 cm). Obviously FLRW(1) can never exceed the irrigation intensity (INTENS).

4.3.2 *Rain*

If rain has to be considered, instead of irrigation, the program has to be modified. The amount of rain can best be included by a RAIN FUNCTION, defining the rain intensity in cm as a function of the TIME, e.g.,

```
FUNCTION RAINTB = 0., 0., 5.,0., 5.01,
0.01, 12., 0.01, 12.01, 0.,  ...
```

This means that there is no rain during the first 5 days, then follow 7 days with a mean rain intensity of 0.01 cm per day, thereafter rain stops, etc. It is convenient to take the rain periods long enough so that, on the average, no ponding occurs. The special program instructions for ponding can than be avoided.
The water flow rate can now be derived with

```
FLRW(1) = AFGEN(RAINTB,TIME)
```

4.4 Leaching of nitrate

In the model leaching of nitrate is caused by diffusion, mass flow and dispersion.
The inflow of nitrate in the top layer (SFLW(1)) depends on the nitrate concentration of the irrigation water (CIR, mg N cm^{-3}) and the flow rate of water (FLRW(1))

```
SFLW(1) = FLRW(1) * CIR                                   (4.36)
```

The flow of nitrate by mass flow (MFL, mg N day^{-1}) for the layers 2–20 is obtained with

```
MFL(I) = FLRW(I) * INSW(FLRW(I),
  CNO(I),CNO(I-1))                                        (4.37)
```

The CSMP routine INSW ensures that FLRW(I) is multiplied by CNO(I-1) if the flow of water is downwards and multiplied by CNO(I) if the flow of water is upwards.
The outflow (SFLW(21)) is obtained with

```
SFLW(21) = FLRW(21) * CNO(20)                    (4.38)
```

The amount of nitrate leached out of the profile (OUTFLS, mg N) is obtained with

```
OUTFLS = INTGRL(0.,SFLW(21))                     (4.39)
```

Diffusion is another process by which nitrate moves in the soil. The diffusion depends on the concentration gradient and the diffusion coefficient. The diffusion coefficient is less in soil than in pure water for two reasons. First when only a part of the soil volume is occupied by water, diffusion is restricted to this part. Second, the water filled pores in the soil are not straight capillaries in the vertical direction, but form a labyrinth, so that the diffusion path between two points is longer than the direct distance between them. To obtain the diffusion coefficient in soil, the coefficient in water (DIF, cm^2 day^{-1}) must be multiplied by the water content (WC) and the labyrinth factor (TORT), which is assumed to be uniform for all layers. Both WC and TORT are dimensionless. According to Frissel et al. (1970) this labyrinth factor depends on the water content, its value is less than one.
So the diffusion constant of nitrate in soil (cm^2 day^{-1}) equals DIF * TORT * (0.5 × (WC(I-1) + WC(I))), and subsequently the flow of nitrate by diffusion equals DIF * TORT * (0.5 * (WC(I-1) + WC(I))) * (CNO(I-1) - CNO(I))/DIFD(I), in which CNO represents the concentration of nitrate expressed as mg N cm^{-3}.
The flow of water through the soil pores causes dispersion of the nitrate. This dispersion is, according to Reiniger (1970), proportional to the concentration gradient and the absolute flow rate of water, ABS(FLRW(I)).
So the nitrate flux by dispersion (mg N day^{-1}) equals:

```
ABS(FLRW(I))*DISP*(CON(I-1)
-CNO(I))/DIFD(I),
```

in which the dispersion coefficient (DISP, cm) is assumed to be uniform for all layers.

Since the mass flow equations are applied to soil layers of finite thickness an error, which causes a dispersion effect, is introduced. This effect can be corrected by the following equation,

```
-ABS(FLRW(I))*(DIFD(I)/2.)*(CNO(I-1)
-CNO(I))/DIFD(I):
```

which in fact causes a negative dispersion effect. The integrated effect of the diffusion process (DIFFL, mg N day^{-1}) for the layers 2–20 is calculated in DO-loop 21.

```
DO 21 I = 2,20
APDIF(I) = DISP * ABS(FLRW(I)) +
  DIF * TORT * (0.5 * (WC(I-1) +
  WC(I))) - ABS(FLRW(I)) *
  DIFD(I)/2.                                  (4.36)
DIFFL(I) = APDIF(I) * (CNO(I-1)-
  CNO(I))/DIFD(I)                             (4.40)
```

The value of the dispersion coefficient (DISP) is about 0.7 for course sand and up to 7 cm for löss, according to Frissel et al. (1970). In the model, DISP is taken as equal to 2 cm.

The diffusion coefficient of nitrate in pure water (DIF) is set to 1 cm^2 day^{-1}. The labyrinth factor has been taken as equal to 0.6 for all layers, so TORT = 0.6.

The nitrate flow (SFLW, mg N day^{-1}) caused by three processes is defined by

```
SFLW(I) = DIFFL(I) + MFL(I)                   (4.41)
21 CONTINUE
```

The net flow of nitrate (NFNO, mg N day^{-1}) depends for the layers 1–4 not only on the inflow and outflow of nitrate but also on the rate of nitrification (RNITF, mg N day^{-1}) and the consumption rate of nitrate (CRNO, mg N day^{-1}) due to the decomposition of organic material with a high C/N quotient

```
DO 23 I = 1,4
NFNO(I) = SFLW(I) - SFLW(I + 1) +
  RNITF(I) + CRNO(I)                          (4.42)
23 CONTINUE
```

The net flow (NFNO) for the layers 5–20 is obtained with

```
DO 24 I = 5,20
NFNO(I) = SFLW(I) - SFLW(I+1)                    (4.43)
24 CONTINUE
```

The amount of nitrate per layer (NO3B, mg N) is now obtained with

```
NO3B1 = INTGRL(INOB1, NFNO1,20)                  (4.44)
```

in which the initial amount of nitrate per layer (INOB, mg N) is calculated from the initial concentration of nitrate (ICNO3, mg N cm^{-3}) and the initial volume of water (IVLW, cm^3 The).nitrate concentration for each layer (CNO, mg N cm^{-3}) is calculated from the amount of nitrate (NO3B) and the volume of water per layer (VOLW).

```
DO 8 I = 1,20
CNO(I) = NO3B(I)/VOLW(I)                         (4.45)
8 CONTINUE
```

5 General features of the simulation model

In the preceding chapters the separate processes which occur in the soil have been described, now certain points concerning the simultaneous action of the various processes in the model as a whole will be discussed.

5.1 Programming

Since the soil is divided in compartments or layers (Fig. 2) indexed variables (arrays) have been used for the definition of the different processes. Arrays are defined by the statements /REAL,/ EQUIVALENCE and STORAGE. In the CSMP program, which is added as an appendix, these statements are indicated by the abbreviation STORE to improve readability. Statements used to define decomposition processes are indicated by the abbreviation DEC, for the nitrification process by NITR, for the flow of heat by TEMP, for the flow of water by WAT, and for the flow of nitrate by SALT. The statements which define the geometry of the soil column are indicated by GEOM.

Integration is performed by the Milne fifth order predictor-corrector method with a variable time step, as stated by the statement METHOD MILNE (see User's Manual, IBM 1967). Quantities are expressed as mg or cm^3 cm^{-2} of surface area per layer thickness. For practical reasons these units are converted to kg or m^3 per hectare per layer thickness by multiplying the former units by a factor (FAC2) equal to 100.

Statements providing these conversion are indicated by CONV in the program.

The results of the computation are presented in three different forms by the statements, PRINT, WRITE, PRTPLOT. The PRTPLOT output is given daily, (OUTDEL = 1.), the PRINT output is given every two days (PRDEL = 2.). The output specified by WRITE is also presented at two-day intervals. This is not a standard output and is realized by:

```
A = IMPULS(0.,PRDEL)
IF (A * KEEP .LT. 0.5) GO TO 30
WRITE (...
30 CONTINUE
```

The subroutine DEBUG may be called by the statements:

```
IF(KEEP .LT. 0.5) GO TO 29
Y = DEBUG(5,0.)
29 CONTINUE
```

The DEBUG routine is very useful in the problem checkout phase. Further information about this subroutine is available in the User's manual, IBM (1967).

Since the present simulation model is still in its developing phase provision has been made that no computation time will be wasted by errors. So a terminating statement FINISH TELLER=*n* is included. This statement counts the number of times that a certain calculation is performed, if TELLER has reached a value of *n*, the simulation is halted. In the presented example no errors were expected and *n* was set to 13000.

5.2 Discussion of the results

The results of the simulation of the discribed example are given in Figs 8–14. As explained a 45-day period was simulated, and on the morning of the 6th day 7.5 cm irrigation water was supplied. Fig. 8 shows the water profile after 4, 6, 8 and 44 days. During the first 4 days the water content in the column stays very close to 0.15 for each layer; at the end of the 6th day the water front has already reached a depth of 66 cm; by the 8th day the irrigated water is more or less evenly distributed over the profile with a water content of about 0.22. After day 8 the changes are small, in the next 36 days the water content returns to about 0.175.

The temperature curves are shown in Fig. 9.

Starting at 10 °C the soil warms up rather fast and after 4 days almost reaches its final value. The temperature of the irrigation water is, however, 16 °C and disturbes this equilibrium. It takes again 2 days to reach equilibrium. Note that this is a dynamic equilibrium. In Fig. 9

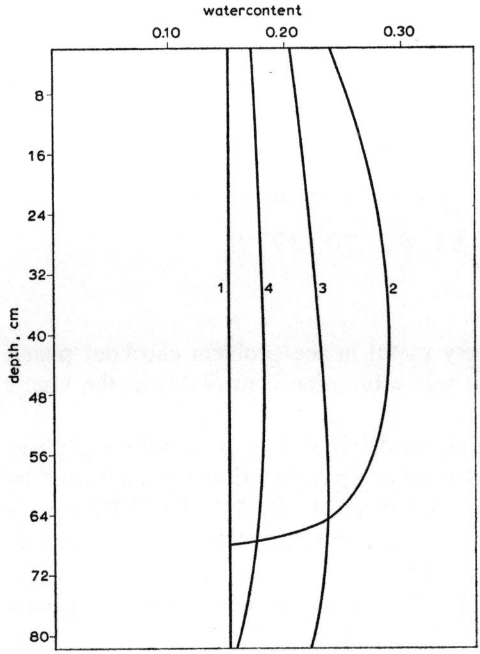

Fig. 8 | Water content as a function of depth and time. Curve 1: 4 days, Curve 2: 6 days, Curve 3: 8 days, Curve 4: 44 days.

only the midnight temperature is presented, in fact the temperature varies daily between 10 and 21 °C owing to the variation of the air temperature between 10° and 22 °C.

The decomposition of organic matter is shown in Fig. 10. In 44 days about two thirds of the protein has decomposed, from the sugar less than a quarter of the original amount remains. The decomposition of lignin and cellulose can almost be neglected for such short periods. Biomass is built up and increases from 12 kg to 72 kg per ha per layer. Differences between the 4 layers are small, see Table 4.

The C/N quotient of lignin is shown in Fig. 11, in a week the C/N quotient decreases from 100 to 50, in three weeks it is down to 30. (Remember that this is a simulation, it is not a proof that lignin takes up nitrogen).

Fig. 12 shows the number of ammonium oxidizers as a function of time for layers 1 and 4. During the first week there is almost no growth, then growth starts with an exponential-like curve which is so typical

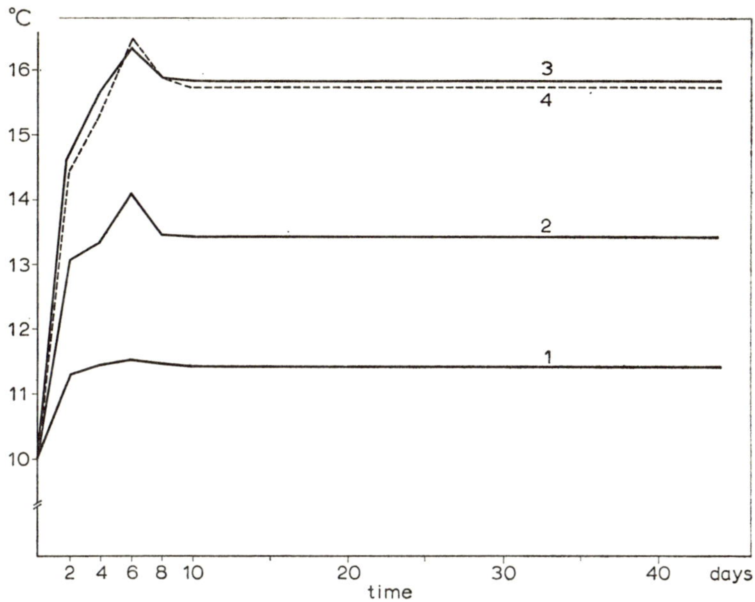

Fig. 9 | Temperature as a function of depth and time. Curve 1: depth 2 cm, Curve 2: depth 6 cm, Curve 3: depth 18 cm, Curve 4: depth 38 cm.

Table 4 The composition of organic matter after 44 days of decomposition.

Organic matter, kg per ha per 4 cm layer					
layer	protein	sugar	lignin	cellulose	biomass
1	28	108	307	398	72
2	28	106	307	398	72
3	27	106	307	398	72
4	27	105	307	397	72
Original composition per layer	116	312	312	427	12

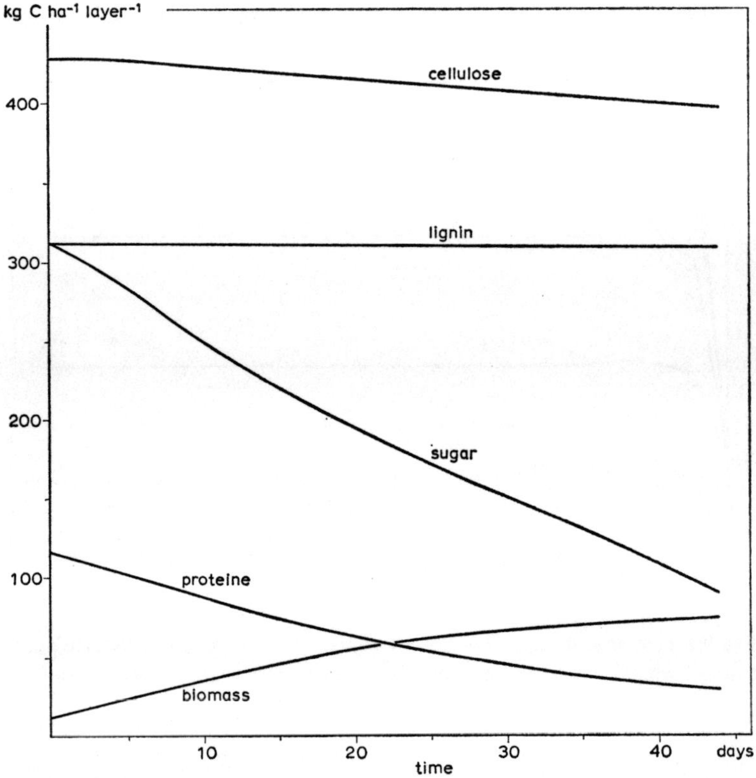

Fig. 10 | Organic material, kg per ha for the upper 4 cm layer, as a function of time.

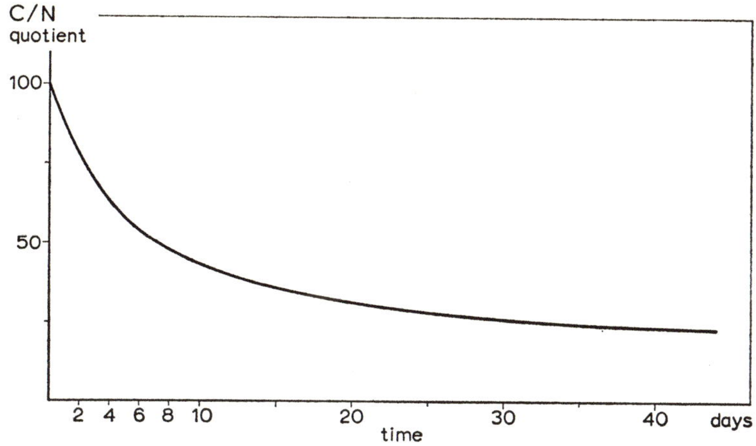

Fig. 11 | C/N quotient of lignin as a function of time.

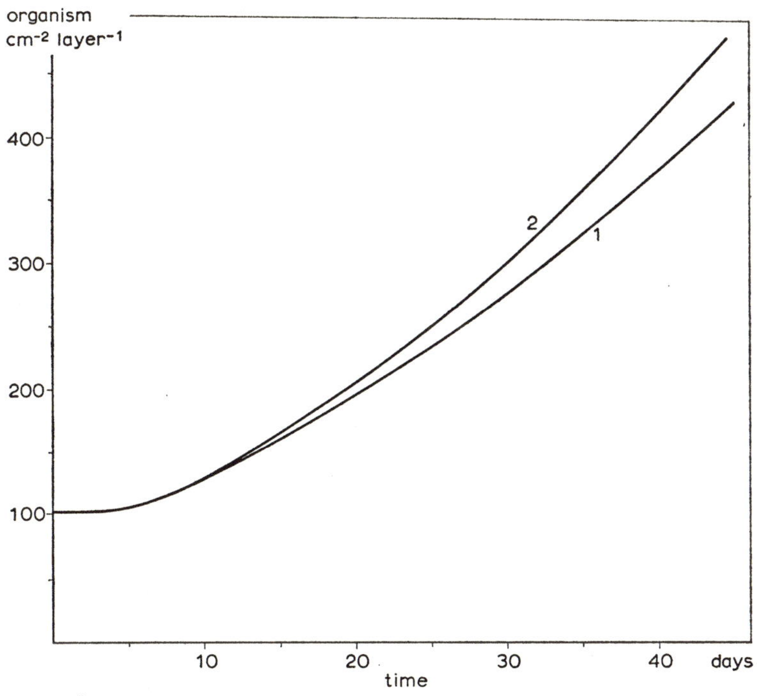

Fig. 12 | Number of micro-organisms oxidizing ammonia, as a function of time and depth. Curve 1: depth 2 cm, Curve 2: depth 14 cm.

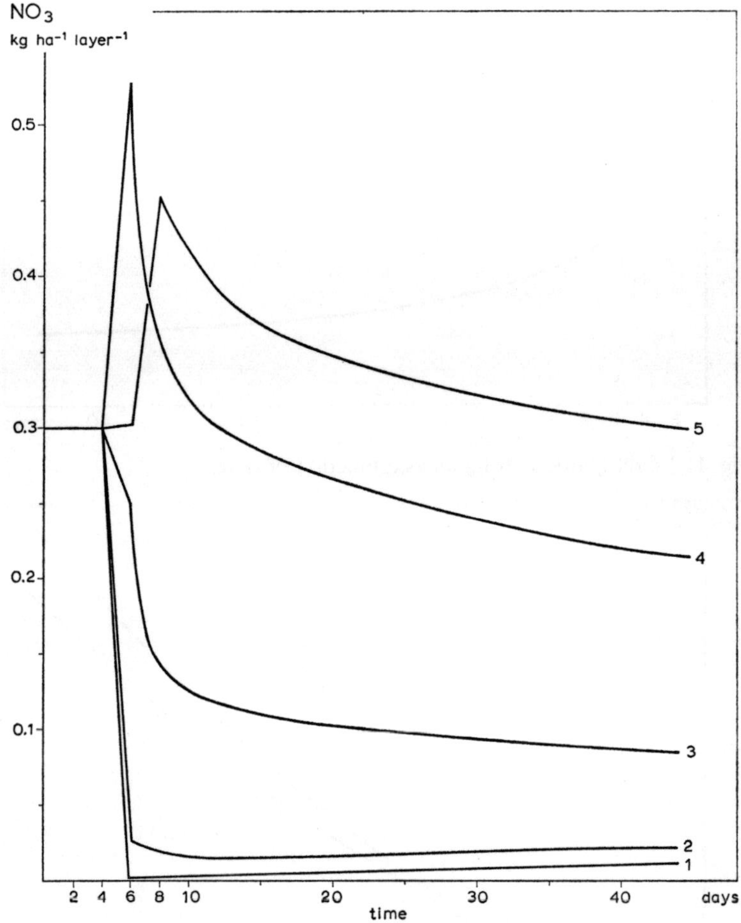

Fig. 13 | Nitrogen as nitrate in kg per ha per 4 cm layer as function of time and depth. Curve 1: depth 2 cm, Curve 2: depth 10 cm, Curve 3: depth 26 cm, Curve 4: depth 46 cm, Curve 5: depth 70 cm.

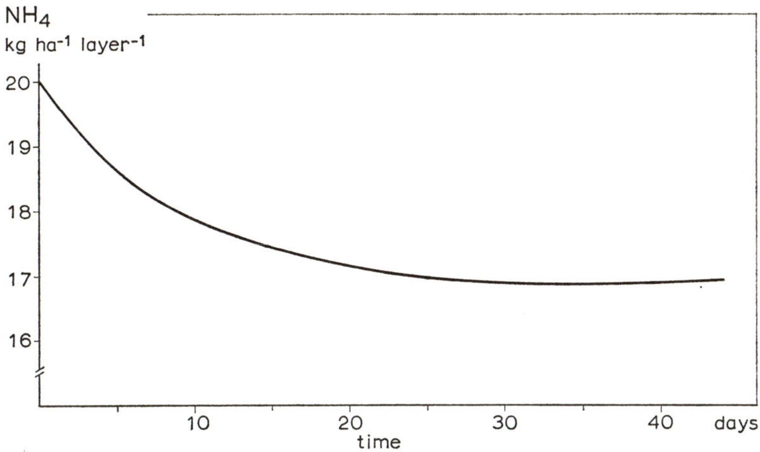

Fig. 14 | Nitrogen as ammonium in kg per ha for the upper 4 cm layer as function of time. The curves for the layers at 6, 10 and 14 cm depth are almost identical.

for microbial processes. Conditions in the fourth layer are clearly better than in the upper layer.

The leaching of the nitrate-N is shown in Fig. 13. The first 5 days there is no leaching. (Because day 5 is not printed it seems as if leaching already starts at day 4. This is of course not true, see Fig. 15 which is an example of the PRTPLOT output that uses daily intervals). Layer 1 is leached out extremely fast, but the third layer (Curve 2, depth 10 cm) also looses its nitrate-N quickly. After 2 weeks, when the water migration has almost finished, the nitrate-N content increases again due to the decomposition of the organic matter. The leaching of Layer 7 (Curve 3, depth 26 cm) is less pronounced, there is no increase due to decomposition of organic matter.

The layers 7 and 18 (Curve 4, depth 46 cm and Curve 5, depth 70 cm) at first show a considerable increase of their nitrate content. This is of course caused by the higher soil moisture content and not by an increase in nitrate concentration.

The ammonium-N (Fig. 14) is leached out more slowly, the nitrogen content goes back from 20 kg to 17 kg per 4 cm layer per ha. It must be noted, however, that desorption of ammonium is not included in the

```
                              MINIMUM                 WCGM1  VERSUS TIME        MAXIMUM
                              1.4886E-01                                        3.7127E-01
 TIME            WCGM1        I                                                 I
 0.0             1.5000E-01   +
 1.0000E 00      1.4965E-01   +
 2.0000E 00      1.4941E-01   +
 3.0000E 00      1.4921E-01   +
 4.0000E 00      1.4905E-01   +
 5.0000E 00      1.4890E-01   +
 6.0000E 00      2.4332E-01   ----------------------+
 7.0000E 00      2.1266E-01   ---------------+
 8.0000E 00      2.0543E-01   -------------+
 9.0000E 00      2.0138E-01   ------------+
 1.0000E 01      1.9846E-01   ------------+
 1.1000E 01      1.9607E-01   -----------+
 1.2000E 01      1.9408E-01   -----------+
 1.3000E 01      1.9235E-01   ----------+
 1.4000E 01      1.9083E-01   ----------+
 1.5000E 01      1.8947E-01   ----------+
 1.6000E 01      1.8824E-01   ---------+
 1.7000E 01      1.8711E-01   ---------+
 1.8000E 01      1.8607E-01   ---------+
 1.9000E 01      1.8510E-01   ---------+
 2.0000E 01      1.8420E-01   --------+
 2.1000E 01      1.8336E-01   --------+
 2.2000E 01      1.8256E-01   --------+
 2.3000E 01      1.8181E-01   --------+
 2.4000E 01      1.8109E-01   --------+
 2.5000E 01      1.8041E-01   --------+
 2.6000E 01      1.7977E-01   -------+
 2.7000E 01      1.7916E-01   -------+
 2.8000E 01      1.7857E-01   -------+
 2.9000E 01      1.7802E-01   -------+
 3.0000E 01      1.7750E-01   -------+
 3.1000E 01      1.7700E-01   -------+
 3.2000E 01      1.7653E-01   -------+
 3.3000E 01      1.7608E-01   -------+
 3.4000E 01      1.7566E-01   -------+
 3.5000E 01      1.7526E-01   ------+
 3.6000E 01      1.7487E-01   ------+
 3.7000E 01      1.7450E-01   ------+
 3.8000E 01      1.7414E-01   ------+
 3.9000E 01      1.7380E-01   ------+
 4.0000E 01      1.7346E-01   ------+
 4.1000E 01      1.7312E-01   ------+
 4.2000E 01      1.7280E-01   ------+
 4.3000E 01      1.7248E-01   ------+
 4.4000E 01      1.7217E-01   ------+
 4.5000E 01      1.7187E-01   ------+
```

Fig. 15 | Example of a PRTPLOT output. The water content of the first layer (WCGM1) is plotted automatically against the TIME. Note the water content at the axis varies between 1.4886 E–01 and 3.712 E–01. The last value is close to the saturation value which equals 0.43.

programme, as it must first be transferred into nitrate before it can be leached out. Ammonium-N increases at the end of the simulated period. The differences between the 4 layers are small. After 44 days the amouts in the 4 layers are respectively, 17.0, 17.0, 17.1 and 17.1 kg N per ha per 4 cm layer.

As has been stated, the values of the parameters had to be approximated from different experiments. Consequently, it was not possible to compare the calculated results, according to the integrated action of the separate processes, with these experimental data. The authors are not aware of experiments that are suitable for the experimental verification of the developed model as a whole.

5.3 Further studies

The present model needs about 0.25 minute of computation time to simulate one day. This computation time has to be reduced, either by simplifying some of the time-consuming processes or by accepting a less accurate integration routine.
From other studies it is known that the calculation of the water migration after irrigation is a particularly time-consuming process. So it seems worthwhile, if there is no special reason to maintain the water migration within the program, to calculate the migration of water separately or to use measured data.
If only rain is considered and not irrigation the advantages are somewhat less pronounced.
A simpler calculation of the water migration in a soil is to be described by Makkink & van Heemst and published in 1973 (1973.)
An example of the direct use of graphs or tables describing the water migration will be described by Frissel (1973).
Another calculation that uses much computer time is the daily temperature cycle. The influence of the temperature on microbial processes is very important so that an exact calculation may be worthwhile for the first days and during the irrigation. As after 14 days the temperature cycle repeats itself each day, it would be better then to replace the temperature calculation by a simpler method, e.g. by tables or by a sinus generator for each of the 4 upper layers.

Although it seems hardly credible it is necessary to carry out several experiments to find the missing input and output data. Only then can the program as a whole be tested by comparing the calculated results with the experimental ones.

Before more complex soil systems can be simulated it will be necessary to include the microbial processes of denitrification and of fixation of air nitrogen in the model. Provision has also to be made in the model for the influence of the soil pH on the microbial processes and the influence of the latter processes on the pH. Besides this, uptake of inorganic nitrogen by plants has to be taken into consideration.

References

Alexander, M., 1961. Introduction to soil microbiology. J. Wiley Sons, N.Y.

Anonymous: CSMP User's Manual. IBM Newsletter H20–0367–2.

Duin, R. H. A. van, 1956. Over de invloed van grondbewerking op het transport van warmte, lucht en water in de grond. Versl. landbouwk. Onderz. 62.7.

Flaig, W. 1967. Landbauforschung Völkenrode, vol. 17.

Frissel, M. J., P. Poelstra & P. Reiniger, 1970. Chromatographic transport through soils. Pl. Soil 33: 161–176.

Frissel, M. J. (Ed.), 1973. Simulation of accumulation and leaching in soils, Pudoc, Wageningen.

Greaves, I. E. & E. Carter, 1920. Influence of moisture on the bacterial activities of the soil. Soil Sci. 10: 361.

Harmsen, G. W. & G. J. Kolenbrander, 1965. Soil inorganic nitrogen, Rapport 13, Instituut voor Bodemvruchtbaarheid, Groningen, The Netherlands.

Kolenbrander, G. J., 1969. Personal communication.

Kononova, M. M., 1961. Soil organic matter, 2nd English edition. Pergamon Press.

Makkink, G. F. & H. D. J. van Heemst, 1973. Simulation of the terms of the waterbalance in a soil. Pudoc, Wageningen.

Mayaudon, J. and L. Batistic, 1970. Dégradation biologique de la lignine ^{14}C dans le sol. Ann. Inst. Pasteur 118: 191–198.

Mayaudon, J. & P. Simonart, 1963. Humification des microorganismes marqués par ^{14}C dans le sol. Ann. Inst. Pasteur 105: 257–266.

Quastel, J. H. & P. G. Scholefield, 1951. Biochemistry of nitrification in soil. Bact. Rev. 15: 1–53.

Reiniger, P., 1970. Movement and exchange of sodium and calcium in calcareous and gypseous soils. A computer approach to soil chromatography. Thesis, Hebrew University, Israel.

Reiniger, P., D. H. Yaalon & M. J. Frissel, 1972. Simulation of cation exchange and movement in soils, published in "Isotopes and Radiation in Soil-Plant Relationships including Forestry", IAEA, Vienna.

Russel, J. C., E. G. Jones & G. M. Bahrt, 1925. The temperature and moisture factors in nitrate production. Soil Sci. vol. 19.

Schreven, D. A. van, 1964. A comparison between the effect of fresh and dried organic materials added to soil on carbon and nitrogen mineralization. Pl. Soil 20 no. 2.

Staple, W. J., 1969. Comparison of computed and measured moisture redistribution following infiltration. Soil Sci. Soc. Amer. Proc. 33: 6.

Stojanovic, B. J. & M. Alexander, 1958. Effect of inorganic nitrogen on nitrification. Soil Sci. vol. 86.

Wit, C. T. de & H. van Keulen (Ed.), 1972. Simulation of transport processes in soils. Pudoc, Wageningen.

Appendix

```
     ****CONTINUOUS SYSTEM MODELING PROGRAM****

  ***PROBLEM INPUT STATEMENTS***

TITLE DECOMPOSITION ORGANIC MATTER, NITRIFICATION OF NH4 AND LEACHING OF NO3
********************************************************************************

*                            EURATOM-ITAL P.O. BOX 48 WAGENIGEN NETHERLANDS

*DIMENSIONS  MG, CM, DAY, DEGREE CENTIGRADE,CAL

* MACRO FOR DECOMPOSITION RATE

MACRO RDECX=DEC1(CMPX,SPERA,SPRCFX,KX,SX,WDEC,TDEC,CNBIMA,EFX,CNCMX,NIT)DEC

      RDECX=CMPX*SPERA*SPRCFX*KX*SX*WDEC*TDEC                          DEC
      KX=INSW((CNBIMA/EFX-CNCMX),FCNSW(NIT,0.,0.,1.),1.)              DEC
      SX=FCNSW(CMPX,0.,0.,1.)                                         DEC
ENDMAC                                                                DEC

* MACRO FOR PRODUCTION RATE

MACRO GRBIMX,LRCO2X,RPRN3X,RCON3X,RCONOX=DEC2(PX,EFX,RDECX,CNCMX,...  DEC
      CNBIMA,BNH3,BNO3)                                               DEC
      GRBIMX=EFX*RDECX                                                DEC
      LRCO2X=(1.-EFX)*RDECX                                           DEC
      PX=1./CNCMX-EFX/CNBIMA                                          DEC
      RPRN3X=RDECX*INSW(PX,0.,PX)                                     DEC
      RCON3X=RDECX*FCNSW(PX,PX,0.,0.)*INSW(BNH3,0.,1.)                DEC
      RCONOX=RDECX*FCNSW(BNH3,1.,0.,0.)*FCNSW(PX,PX,0.,0.)            DEC
ENDMAC                                                                DEC

*****************************************************************************

INITIAL
NOSORT

PARAMETER N=20                                                        GEOM
*         N= NUMBER OF SOIL LAYERS
             TELLER = 0.
FIXED     I,N                                                         GEOM

/      REAL VOLW(25),IVLW(25),NFLW(25),NO3B(25),INOB(25),NFNO(25),T(25)  STORE
/      REAL HCL(15),IHCL(15),NHFL(15),COMP(50),ICMP(50),RDC(50),VOLL(25) STORE
/      REAL AMOX(10),IAMOX(10),NGRAM(10),MRNH3(10),WC(25),D(25)         STORE
/      REAL NH3B(10),INH3B(8),NFNH3(8),TOTCOM(5),DIFD(25),HNL(4)        STORE

STORAGE CNO(25),AVDIF(25),AVCOND(25),FLRW(25),HEFLW(15)                 STORE
STORAGE APDIF(25),DIFFL(25),MFL(25),SFLW(25)   ,DEPTH(25)               STORE
STORAGE COND(25),CRNO(10),RNITF(10),HECO(15),AVHECO(15)                 STORE
STORAGE HEDI(15),PCOMP(5),SCOMP(5),LCOMP(5),CCOMP(5)                    STORE
STORAGE BCOMP(5),NITRAT(25),AMMON(5),CNL(4)                             STORE

/      EQUIVALENCE (VOLW1,VOLW(1)),(IVLW1,IVLW(1)),(NFLW1,NFLW(1))      STORE
/      EQUIVALENCE (NO3B1,NO3B(1)),(INOB1,INOB(1)),(NFNO1,NFNO(1))      STORE
/      EQUIVALENCE (HCL1,HCL(1)),(IHCL1,IHCL(1)),(NHFL1,NHFL(1))        STORE
/      EQUIVALENCE (COMP1,COMP(1)),(ICMP1,ICMP(1)),(RDC1,RDC(1))        STORE
/      EQUIVALENCE (AMOX1,AMOX(1)),(IAMOX1,IAMOX(1)),(NGRAM1,NGRAM(1))  STORE
/      EQUIVALENCE (NH3B1,NH3B(1)),(INH3B1,INH3B(1)),(NFNH31,NFNH3(1))  STORE

* GEOMETRY OF THE SYSTEM  20 LAYERS

      AREA=1.                                                         GEOM
*         AREA = AREA OF SOIL LAYER TREATED IN CM**2
*         TCOM= THICKNESS OF LAYERS IN CM
STORAGE TCOM(20)                                                      GEOM
TABLE TCOM(1-20)=20 *4.                                               GEOM
*         DIFD= DISTANCE BETWEEN CENTERS OF  2 LAYERS IN CM
```

```
*        DEPTH=DISTANCE FROM SURFACE TO CENTER OF LAYER,CM
*        VOLL=VOLUME OF SOIL LAYER IN CM**3
            DIFD(1)=0.5*TCOM(1)                                    GEOM
            DEPTH(1)=0.5*TCOM(1)                                   GEOM
      VOLL(1)=AREA*TCOM(1)                                         GEOM

      DO 1 I=2,N                                                   GEOM
            DIFD(I)=0.5*(TCOM(I)+TCOM(I-1))                        GEOM
            DEPTH(I)=DEPTH(I-1)+DIFD(I)                            GEOM
      VOLL(I)=AREA*TCOM(I)                                         GEOM
 1    CONTINUE                                                     GEOM

* INITIAL SITUATION OF WATER AND NO3*************************************

      IWC=0.15                                                     WAT
      ICNO3=0.5E-2                                                 SALT
* IWC = INITIAL WATER CONTENT

* ICNO3 = INITIAL CONC. OF NO3, MG N PER CM**3 WATER
      DO 2 I=1,N                                                   WAT
      IVLW(I)=VOLL(I)*IWC                                          WAT
      INOB(I)=ICNO3*IVLW(I)                                        SALT
 2    CONTINUE                                                     WAT

* INITIAL SITUATION OF TEMPERATURE**************

PARAMETER HECAM=0.48                                               TEMP
PARAMETER HECAW=1.                                                 TEMP
PARAMETER INTEMP=10.                                               TEMP
PARAMETER AMMIN=0.57                                               TEMP

         HECAMI=AMMIN*HECAM                                        TEMP
         IHECA=HECAMI+IWC    *HECAW                                TEMP

      DO 3 I=1,10                                                  TEMP
      IHCL(I)=IHECA*VOLL(I)*INTEMP                                 TEMP
 3    CONTINUE                                                     TEMP

*        INTEMP=INITIAL TEMPERATURE OF ALL LAYERS, DEGREE CENTIGRADE
*        AMMIN=RELATIVE AMOUNT OF MINERALS IN SOIL VOLUME
*        HECAM=HEAT CAPACITY,MINERAL PART,CAL/CM**3*DEGR
*        HECAMI=HEAT CAPACITY OF MINERALS PER CM**3 OF SOIL
*        HECAW=HEAT CAPACITY WATER, CAL/CM**3*DEGREE
*        IHECA=INITIAL HEAT CAPACITY OF SOIL VOLUME,CAL/CM**3*DEGR
*        IHCL  =INITIAL HEAT CAPACITY OF WHOLE LAYER,CAL

* INITIAL AMOUNTS   OF COMPONENTS OF ORGANIC MATTER AND NH3*************

* COMPONENTS ARE EXPRESSED IN MG C/CM**3 SOIL,NH4 IN MG N/CM**3 SOIL
* THIS DOLOOP CALCULATES THE INITIAL AMOUNT OF EVERY COMPONENT
* FOR ALL LAYERS

      DO 4 I=1,4                                                   DEC
      ICMP(I)=PROT*VOLL(I)                                         DEC
      ICMP(I+4 ) =SUGAR*VOLL(I)                                    DEC
      ICMP(I+8 )=LIGN*VOLL(I)                                      DEC
       IAMOX(I+4)=ICMP(I+8)/100.                                   DEC
      ICMP(I+12)=CELLU*VOLL(I)                                     DEC
 4    ICMP(I+16)=BIMA*VOLL(I)                                      DEC

* IAMOX5 = INITIAL AMOUNT OF ORGANIC NITROGEN IN LIGNIN, MG N PER LAYER
INCON PROT=0.29,SUGAR=0.78,CELLU=1.07,LIGN=0.78,BIMA=0.03          DEC
* PROTEINS, SUGARS, CELLULOSE, LIGNIN, BIOMASS, MG C PER CM**3 SOIL

INCON AMNH3=0.05                                                   DEC
* INITIAL AMOUNT OF AMMONIUM PER CM**3 SOIL  VOLUME

      DO 5 I=1,4                                                   DEC
      INH3B(I)=AMNH3*VOLL(I)                                       DEC
 5    CONTINUE                                                     DEC
```

```
* INITIAL SITUATICN OF AMMONIUM OXYDIZERS*****************

      INAM=26.                                                         NITR
* INAM IS NUMBER OF AMM. OXYDIZERS PER CM**3 SOIL AT TIME=0
* MRNH3N = MAX. RATE OF NITRIFICATION, MG N / CM**3 SOIL PER DAY

*          BULK DENSITY=1.3
*        CONVERSION OF MICROGR N PER G SOIL TO MG PER CM**3 SOIL
      MRNH3N=70.*1.3/1000.                                             NITR


* MRNH3 = MAX. RATE OF NITRIFICATION, MG N PER DAY PER SOIL LAYER
      DO 6 I=1,4                                                       NITR
      MRNH3(I)=MRNH3N*VOLL(I)                                          NITR
 6    CONTINUE                                                         NITR

      DO 7 I=1,4                                                       NITR
      IAMOX(I)=INAM*VOLL(I)                                            NITR
 7    CONTINUE                                                         NITR
****************************************************************************

DYNAMIC
NOSORT
            TELLER = TELLER + 1.

*           DO LOOP FOR CALCULATIONS OF SOME VARIABLES DEPENDING ON
*           WATERCONTENT

      DO 8 I=1,N                                                       WAT
            WC(I) = VOLW(I)/VOLL(I)                                    WAT
      D(I)=AFGEN(DTBL,WC(I))                                           WAT
            COND(I) = AFGEN(CONDTB,WC(I))                              WAT
      CNC(I)=NO3B(I)/VOLW(I)                                           SALT
 8    CONTINUE                                                         WAT

* WC IS WATER CONTENT OF THE LAYERS
* CNO IS CONCENTRATION NO3 OF THE LAYERS IN MG N/CM**3 WATER

*          COND=CONDUCTIVITY OF SOIL TO WATER IN CM/DAY

*          D=DIFFUSIVITY, CM**2/DAY
*             CONDUCTUVITY AND DIFFUSVITY TABLE IN CM/DAY RESP CM**2/DAY
*             AS FUNCTION OF WATERCONTENT

FUNCTION    DTBL=          ...                                         WAT
            (0.02,4.),(0.05,1.) ,(0.07,1.),(0.14,5.), ...              WAT
            (0.157,10.),(0.20,50.),(0.22,100.),(0.267,500.),  ...      WAT
            (0.290,1000.),(0.35,5000.),(0.38,10000.),(0.4300,6.E4)     WAT

FUNCTION    CONDTB =                    ...                            WAT
            (0.018,1.E-8),(0.03,1.E-7),(0.057,1.E-6),(0.075,5.E-6),  ...WAT
            (0.081,1.E-5),(0.108,5.E-5),(0.114,1.E-4),(0.136,5.E-4), ...WAT
            (0.147,0.001),(0.164,0.005),(0.175,0.01),(0.202,0.05),   ...WAT
            (0.212,0.1),(0.234,0.5),(0.249,1.),(0.30,5.),(0.342,10.),...WAT
            (0.43,30.),(0.431,30.)                                     WAT

*          DSRF=DIFFUSIVITY OF SURFACE LAYER
*          CONDSF=CONDUCTIVITY OF SURFACE LAYER
*          DSL= DIFFUSIVITY IN LOWER BOUNDARY
*          CONDL=CONDUCTIVITY IN LOWER BOUNDARY

            DSRF = AFGEN(DTBL,WCSAT)                                   WAT
            CONDSF = AFGEN(CONDTB,WCSAT)                               WAT
         DSL=AFGEN(DTBL,WCLAST)                                        WAT
         CONDL=AFGEN(CONDTB,WCLAST)                                    WAT

PARAMETER WCLAST=0.15                                                  WAT
PARAM     WCSAT=0.43                                                   WAT
*          WCLAST =WATER CONTENT OF LOWER BOUNDARY,N+1
*          WCSAT=WATER CONTENT SATURATED SAMPLE
```

```
* CALCULATION OF TEMP. OF FIRST 10 LAYERS***************

      DO 9 I=1,10                                                      TEMP
      T(I)=HCL(I)/((HECAMI+WC(I)*HECAW)*VOLL(I))                        TEMP
 9    CONTINUE                                                          TEMP

* T=TEMPERATURE,HCL=HEAT CAP. OF LAYER (CAL)

* CALCULATION OF FLOW RATES OF WATER OVER BOUNDARIES 2-20, CM/DAY

      DO 10 I=2,N                                                       WAT
            AVDIF(I)=(W*D(I-1)+D(I))/(W+1.)                             WAT
            AVCOND(I)=(W*COND(I-1)+COND(I))/(W+1.)                      WAT
            FLRW(I)=AVDIF(I)*(WC(I-1)-WC(I))/DIFD(I)+AVCOND(I)          WAT
 10   CONTINUE                                                          WAT

*        AVDIF=AVERAGE DIFFUSIVITY
*        AVCOND=AVERAGE CONDUCTIVITY

*        FLRW=FLOW RATE OF WATER      CM/DAY

PARAMETER   W=1.                                                        WAT
* W=WEIGHT FACTOR

*FLOW RATE OF WATER OVER LAST BOUNDERY (21)

         AVDIF(N+1)=(W*D(N)+DSL)/(W+1.)                                 WAT
         AVCOND(N+1)=(W*COND(N)+CONDL)/(W+1.)                           WAT
         FLRW(N+1)=AVDIF(N+1)*(WC(N)-WCLAST)/(0.5*TCOM(N))+AVCOND(N+1)  WAT
         FLRW(N+1)=INSW(FLRW(N+1),0.,FLRW(N+1))                         WAT

*FLOW RATE OF WATER OVER FIRST BOUNDERY (1) :BY IRRIGATION

PARAMETER PIRR=14. ,ON=5.33 ,HMAX=0.3 ,WAIR=7.5 ,INTENS=1.              WAT

*        PIRR=IRRIGATION FREQUENCY IN DAYS
*        WAIR= AMOUNT OF WATER APPLIED PER IRRIGATION IN CM
*        INTENS=INTENSITY OF IRRIGATION IN CM/HOUR
*        ON= BEGINNINING OF IRRIGATION IN DAYS
*        DIRR= IRRIGATION TIME DURATION IN DAYS
*        HMAX= MAXIMUM PERMITTED HYDRAULIC HEAD,CM
*        IRR IS IRRIGATION RATE, CM PER DAY

IF(TIME .LT. 14.) GO TO 101
HOFF = 0.
FLRW(1) = 0.
IRR = 0.
GO TO 12

101   QQ = AMOD(TIME, PIRR)
        DIRR = WAIR/(INTENS*24.)
*                                         DIRR = 0.31 FOR THIS CASE
        OFF=ON+DIRR                                                     WAT
IF (QQ .GT. ON) GO TO 11
           HOFF=.0                                                      WAT
           FLRW(1)=0.                                                   WAT
           IRR=.0                                                       WAT
           GOTO 12                                                      WAT

 11        CONTINUE                                                     WAT
           IF(QQ  .GT. OFF) GO TO 13
           IF(HOFF.GT.0.5) GOTO 13                                      WAT
           IF(H.GT.HMAX) GOTO 14                                        WAT
           AVDIF(1) = (W*DSRF+D(1))/(1.+W)                              WAT
           AVCOND(1) =(W*CONDSF+COND(1))/(1.+W)                         WAT
           INRATE  = AVDIF(1)*(WCSAT-WC(1))/DIFD(1)+AVCOND(1)           WAT
           IRR=INTENS*24.                                               WAT
           IF(IRR.GT.INRATE) GOTO 15                                    WAT
           FLRW(1)=IRR                                                  WAT
           GOTO 12                                                      WAT

 15        CONTINUE                                                     WAT
```

```
            FLRW(1)=INRATE                                                    WAT
            IH=IRR-INRATE                                                     WAT
            GOTO 12                                                           WAT

  13        CONTINUE                                                          WAT
            FLRW(1)=0.                                                        WAT
            IRR=.0                                                            WAT
            IF(H)12,12,16                                                     WAT

  14        HOFF=1.                                                           WAT
            IRR=.0                                                            WAT

  16        CONTINUE                                                          WAT
            AVDIF(1) = (W*DSRF+D(1))/(1.+W)                                   WAT
            AVCOND(1) =(W*CONDSF+COND(1))/(1.+W)                              WAT
            FLRW(1) = AVDIF(1)*(WCSAT-WC(1))/DIFD(1)+AVCOND(1)                WAT
            IH=-FLRW(1)                                                       WAT

  12        LFLR=FLRW(1)                                                      WAT

* CALCULATION OF HEAT FLOW OVER BOUNDARIES 1-11 BY WATER FLOW

*         HEFLW=HEAT FLOW WITH WATER,CAL/DAY*CM**2

      DO 17  I=2,1C                                                           TEMP
         HEFLW(I)=FLRW(I)*HECAW*INSW(FLRW(I),T(I),T(I-1))                     TEMP
   17 CONTINUE                                                                TEMP

         HEFLW(11)=FLRW(11)*HECAW*T(10)                                       TEMP
         HEFLW(1)=FLRW(1)*HECAW*TW                                            TEMP

PARAMETER TW=16.                                                              TEMP

* CALCULATION OF HEAT FLOW BY DIFFUSION OVER BOUNCARY 1-11

*         HECO=HEAT CONDUCTIVITY OF SOIL AT VARIOUS MOISTURE CONTENTS...
*        MCAL/CM*SEC*DEGR
*         HECOF=TABULATED FUNCTION OF HECO DEPENDING ON WATER CONTENT(WC)
*         HECO*24.*3.6 = CAL/CM*DAY*DEGR
*         HEDI=HEAT DIFFUSION THROUGH SOIL, CAL/DAY*CM**2

      DO 18 I=1,11                                                            TEMP
         HECO(I)=AFGEN(HECOF,WC(I))*24.*3.6                                   TEMP
   18  CONTINUE                                                               TEMP

FUNCTION HECOF=0.,0.5,0.05,0.6,0.1,2.1,0.15,2.6,0.2,2.8,...                   TEMP
         0.25,3.1,0.3,3.3,0.4,3.6,0.5,3.9,0.6,4.0                             TEMP

      DO 19 I=2,11                                                            TEMP
         AVHECO(I)=(TCOM(I-1)+TCOM(I))/(TCOM(I-1)/HECO(I-1)+     ...          TEMP
         TCOM(I)/HECO(I))                                                     TEMP
  19  CONTINUE                                                                TEMP

      DO 20 I=2,1C                                                            TEMP
         HEDI(I)=AVHECO(I)*(T(I-1)-T(I))/DIFD(I)                              TEMP
  20  CONTINUE                                                                TEMP

         HEDI(11)=0.                                                          TEMP

* CALCULATION OF HEAT DIF. FLOW FOR FIRST BOUNDERY LAYER

* AIRTF=AIRTEMP VERSUS TIME, GIVEN FOR ONE DAY.

FUNCTION AIRTF=0.,10.,0.25,12.,0.50,20.,0.58,22.,0.75,18.,1.,10.              TEMP
         B=AMOD(TIME,1.)                                                      TEMP
         AIRT=AFGEN(AIRTF,B)                                                  TEMP
         HEDI(1)=HECO(1)*(AIRT-T(1))/DIFD(1)                                  TEMP
```

```
* CALCULATION OF SALT FLOW (NO3) OVER BOUNDARIES I-21

* SALT FLOW IN MG N/DAY
*        TORT=TORTUOSITY FACTOR,ASSSUMED UNIFORM ALL LAYERS
*        DISP=DISPERSION FACTOR ASSUMED UNIFORM ALL LAYERS ,CM
*        DIF=DIFFUSION CONSTANT IN FREE WATER   ,CM2/DAY
*        APDIF=APPARENT DIFFUSION FLOW   ,CM2/DAY
*        DIFFL=DIFFUSION FLOW, MG N PER DAY
*        MFL=MASS FLOW, MG N PER DAY
*        SFLW=SALT FLOW, MG N PER DAY

PARAMETER DISP=2. ,DIF=1. ,TORT=0.6                                       SALT

      DO 21 I=2,N                                                         SALT
            APDIF(I)=DISP   *ABS(FLRW(I))+DIF*TORT  ...                   SALT
            *(0.5*(WC(I-1)+WC(I)))-ABS(FLRW(I))*DIFD(I)/2.                SALT
      DIFFL(I)=APDIF(I)*(CNO(I-1)-CNO(I))/DIFD(I)                         SALT
      MFL(I)=FLRW(I)*INSW(FLRW(I),CNO(I),CNO(I-1))                        SALT
            SFLW(I) = DIFFL(I) + MFL(I)                                   SALT
  21  CONTINUE                                                            SALT

* SALT FLOW OVER FIRST BOUNDARY

            SFLW(1) = FLRW(1)*CIR                                         SALT
*  CIR=CONCENTRATION OF NO3 OF IRRIGATIONWATER, MG N PER CM**3 WATER
PARAMETER CIR=0.                                                          SALT
* SALT FLOW OVER LAST BOUNDARY

      SFLW(N+1)=FLRW(N+1)*CNO(N)                                          SALT


      TLRCO2=0.0                                                          DEC
* TLRCO2=TOTAL LIBERATION RATE OF CO2 OF ALL LAYERS, MG C PER DAY

* DECOMPOSITION AND NITRIFICATION

*     ***SORT-BLOCK***
      DO 22 I=1,4                                                         DEC


********************************************************************************

SORT
      WCL=WC(I)                                                           DEC
      TL=T(I)                                                             DEC
      CMPP=COMP(I)                                                        DEC
      CMPS=COMP(I+4 )                                                     DEC
      CMPL=COMP(I+8 )                                                     DEC
      CMPC=COMP(I+12)                                                     DEC
      CMPB=COMP(I+16)                                                     DEC

*     CMPP=COMPONENT PROTEINS
*     CMPS=COMPONENT SUGARS
*     CMPL=COMPONENT LIGNIN
*     CMPC=COMPONENT CELLULOSE
*     CMPB=COMPONENT BIOMASS



*     BNH3=AMOUNT OF NH4-N OF A LAYER   ,MG N/LAYER
*     BNO3=AMOUNT OF NO3-N   ,MG N/LAYER

      BNH3=NH3B(I)                                                        DEC
      BNO3=NO3B(I)                                                        DEC
      NIT=BNH3+BNO3                                                       DEC

*  AMOXL=NUMBER OF AMMONIUM OXIDIZERS PER SOIL LAYER
*  MRNH3L=MAX. RATE OF NITRIFICATION, MG N PER DAY PER SOIL LAYER

      AMOXL=AMOX(I)                                                       NITR
      MRNH3L=MRNH3(I)                                                     NITR
```

```
*  HNLIG=AMOUNT OF ORGANIC NITROGEN IN LIGNIN, MG N PER SOIL LAYER
        HNLIG=AMOX(I+4)                                                     DEC

* DECOMPOSITION  INFLUENCED BY WATER HOLDING CAPACITY AND TEMP OF THE SOIL
      WHC=WCL/WCSAT *100.                                                   DEC

      WDEC=AFGEN(WDECTB,WHC)                                                DEC

      FUNCTION WDECTB=(0.,0.),(10.,0.02),(30.,0.32),(40.,0.65),...          DEC
      (50.,0.81),(60.,1.),(70.,0.8),(80.,0.55),(90.,0.48),(100.,0.45)       DEC

      FUNCTION  TDECTB=(0.,0.001),(5.,0.15),(10.,0.3),(20.,0.64),...        DEC
                      (30.,0.81),(40.,0.96),(50.,1.)                        DEC

      TDEC=AFGEN(TDECTB,TL)                                                 DEC

*   WHC IS WATER HOLDING CAPACITY OF THE SOIL
*   WDEC=REDUCTIONFACTOR DUE TO PERCENTAGE SATURATION OF WATER HOLDING
*   CAPACITY ON DECOMPOSITION
* WDECTB=TABLE OF WDEC AND WHC(PERCENTAGE SATURATION OF WATER HOLDING
*   CAPACITY)
*   TDEC=REDUCTIONFACTOR DUE TO TEMPERATURE ON DECOMPOSITION
*   TDECTB=TABLE OF TDEC AND TEMP

* MACRO DEC1 AND DEC2 FOR DECOMPOSITION OF THE COMPONENTS

* P REPRESENTS PROTEINS
* S REPRESENTS SUGARS
* L REPRESENTS LIGNIN
* B REPRESENTS BIOMASS
* C REPRESENTS CELLULOSE

*     RDEC=RATE OF DECOMPOSITION MG C/DAY PER LAYER
*     GRBIM=GROWTH RATE OF BIMA    MGC/DAY PER LAYER
*     LRCO2 =  LIBRATION RATE OF CO2  MG C/DAY PER LAYER


* MACROS FOR PROTEINS

RDECP=DEC1(CMPP,SPERA,SPRCFP,KP,SP,WDEC,TDEC,CNBIMA,EFFAC,CNCMP,NIT)     DEC
GRBIMP,LRCO2P,RPRN3P,RCON3P,RCONOP=DEC2(PP,EFFAC,RDECP,CNCMP,CNBIMA,... DEC
BNH3,BNO3)                                                               DEC

PARAMETER SPERA=0.1  ,CNBIMA=8.,EFFAC=0.28,SPRCFP=1.,CNCMP=6.            DEC

* SPERA IS SPECIFIC RATE OF DECCMP. OF PROTEINS, RECIPROCAL DAYS
* CNBIMA IS C/N QUOTIENT OF BIOMASS
* EFFAC IS EFFICIENCYFACTOR
* SPRCFP = SPECIFIC RATE COEFFICIENT FOR PROTEINS
* CNCMP IS C/N-QUOTIENT OF PROTEINS

* MACROS FOR SUGARS

RDECS=DEC1(CMPS,SPERA,SPRCFS,KS,SS,WDEC,TDEC,CNBIMA,EFFAC,CNCMS,NIT)     DEC
GRBIMS,LRCO2S,RPRN3S,RCON3S,RCONOS=DEC2(PS,EFFAC,RDECS,CNCMS,CNBIMA,... DEC
BNH3,BNO3)                                                               DEC

PARAMETER SPRCFS=0.75 ,CNCMS=200.                                        DEC
* MACROS FOR CELLULOSE

RDECC=DEC1(CMPC,SPERA,SPRCFC,KC,SC,WDEC,TDEC,CNBIMA,EFFAC,CNCMC,NIT)     DEC
GRBIMC,LRCO2C,RPRN3C,RCON3C,RCONOC=DEC2(PC,EFFAC,RDECC,CNCMC,CNBIMA,... DEC
BNH3,BNO3 )                                                              DEC

PARAMETER SPRCFC=0.05 ,CNCMC=200.                                        DEC
```

```
* MACROS FOR LIGNIN

RDECL=DEC1(CMPL,SPERA,SPRCFL,KL,SL,WDEC,TDEC,CNBIMA,EFFAC,CNCML,NIT)     DEC
GRBIML,LRCO2L,RPRN3L,RCON3L,RCONOL=DEC2(PL,EFFAC,RDECL,CNCML,CNBIMA,... DEC
BNH3,BNO3)                                                               DEC

PARAMETER  SPRCFL=0.0095                                                 DEC

* Q AND R ARE RATES OF N-FIXATION BY LIGNIN, MG N PER DAY PER SOIL LAYER

         Q=-0.0404*FCNSW(DIFCN,DIFCN,0.,0.)*CMPL*FCNSW(BNH3,0.,0.,1.)    DEC
         R=-0.0404*FCNSW(DIFCN,DIFCN,0.,0.)*CMPL*FCNSW(BNH3,1.,1., ..    DEC
      0.)*FCNSW(BNO3,0.,0.,1.)                                           DEC
         DIFCN=1./CNCML-1./20.                                           DEC
         NRNL=Q+R-RDECL/CNCML                                            DEC
         CNCML=CMPL/HNLIG                                                DEC

* NRNL = NET RATE OF N-FIXATION BY LIGNIN, MG N PER DAY PER LAYER
* CNCML = C/N RATIO OF LIGNIN

*     MACROS FOR BIOMASS

RDECB=DEC1(CMPB,SPERA,SPRCFB,KB,SB,WDEC,TDEC,CNBIMA,EFTOT,CNCMB,NIT)     DEC
GRORMB,LRCO2B,RPRN3B,RCON3B,RCONOB=DEC2(PB,EFTOT,RDECB,CNCMB,CNBIMA,... DEC
BNH3,BNO3)                                                               DEC

* GRORMB = GROWTH RATE ORGANIC MATERIAL DUE TO DECOMPOSITION OF BIOMASS

         REGRBI=GRORMB-EFFHUM*RDECB                                      DEC
         GRHUM=EFFHUM*RDECB                                              DEC
* REGRBI IS REAL GROWTH RATE OF BIOMASS, MG C PER DAY PER LAYER


PARAMETER  SPRCFB=0.5 ,  CNCMB=8.   ,EFFHUM=0.24,EFBIM=0.16,CNHUM=10.    DEC
PARAM  EFTOT=0.40                                                        DEC

* EFFHUM IS EFFICIENCY FACTOR OF HUMUS
* EFBIM IS EFFICIENCY FACTOR OF BIOMASS
* EFTOT IS EFBIM+EFFHUM
* CNHUM = C/N RATIO OF HUMUS
* GRHUM = GROWTH RATE HUMUS ,  MG C PER DAY PER LAYER


      RLCO2L=LRCO2P+LRCO2S+LRCO2C+LRCO2L+LRCO2B                          DEC
*RLCO2L= RATE OF LIBERATION OF CO2 PER LAYER, MG C PER DAY PER LAYER

      NGRBIL=GRBIMP+GRBIML+GRBIMS+GRBIMC+REGRBI-RDECB                    DEC
* NGRBIL = NET GROWTH RATE OF BIOMASS PER LAYER, MG C PER DAY PER LAYER

      PRN3L=RPRN3P+RPRN3S+RPRN3L+RPRN3C+RPRN3B                           DEC
      CRNOL=RCONOP+RCONOS+RCONOL+RCONOC+RCONOB -R                        DEC
      CRN3L=RCON3P+RCON3S+RCON3L+RCON3C+RCON3B -Q                        DEC

*  PRN3L = PRODUCTION RATE OF  NH4-N PER LAYER, MG N PER DAY
*  CRN3L = CONSUMPTION RATE OF NH4-N PER LAYER, MG N PER DAY
*  CRNOL = CONSUMPTION RATE OF NO3-N PER LAYER, MG N PER DAY


*********  NITRIFICATION OF NH4  **********************

         RNH3N=AMIN1(RAMOX*AMOXL*INSW(BNH3,0.,1.),MRNH3L)*TCOF*WCCOF     NITR
         GRAMOX=GRCAO*RNH3N/RAMOX                                        NITR
         DRAMOX=AMOXL/MFAO                                               NITR
         NGRAMX=GRAMOX-DRAMOX                                            NITR

PARAMETER  RAMOX=3.30E-8,GRCAO=0.5,MFAO=30.                              NITR

*    RNH3N=  NITRIFICATION RATE, MG N PER DAY PER SOIL LAYER
*  GRAMOX = GROWTH RATE OF AMMONIUM OXIDIZERS EXPRESSED AS NUMBER/DAY
*  DRAMOX = DEATH RATE AMMONIUM OXIDIZERS PER LAYER, NUMBER/DAY
*   NGRAMX = NET GROWTH RATE OF AM. OX. PER LAYER, NUMBER/DAY
```

```
*                  RAMOX=RELATIVE RATE OF NITRIFICATION OF NH4 EXPRESSED
*                  AS MILLIGRAM NH4-N NITRIFIED PER UNIT OF AMOX PER DAY
*                  GRCAO=GROWTH RATE CONSTANT OF AMOX WITH THE
*                  DIMENSION OF RECIPROCAL DAYS
*                  MFAO=MORTALITY FACTOR OF AMOX EXPRESSED IN DAYS
*  WCCOF = REDUCTION FACTOR DUE TO WHC
*  TCOF = REDUCTION FACTOR DUE TO  TEMP

      WCCOF=AFGEN(WCNFTB,WHC)                                              NITR
      TCOF =AFGEN(TPNFTB,TL )                                              NITR

      FUNCTION  WCNFTB=(0.,0.05),(10.,0.1),(20.,0.17),(30.,0.2),...        NITR
      (35.,0.4),(40.,0.6),(50.,0.85),(60.,1.),(65.,0.95),(70.,0.4),...     NITR
      (80.,0.1),(100.,0.)                                                  NITR

      FUNCTION  TPNFTB=(0.,0.),(10.,0.12),(20.,0.28),(25.,0.4),...         NITR
      (30.,0.85),(35.,1.),(40.,0.95),(45.,0.8),(50.,0.5)                   NITR

*  NFNH3L = NET RATE OF FORMATION OF NH4-N PER LAYER, MG N/DAY
      NFNH3L=PRN3L+CRN3L-RNH3N                                             DEC

NOSCRT

* RATES FROM SORT SECTION HAVE TO BE INDEXED FOR USING THEM IN ARRAY INTEGRALS

      RDC(I)=-RDECP                                                        DEC
      RDC(I+4 )=-RDECS                                                     DEC
      RDC(I+8 )=-RDECL                                                     DEC
      RDC(I+12)=-RDECC                                                     DEC
      RDC(I+16)= NGRBIL                                                    DEC

      RNITF(I)=RNH3N                                                       NITR
      NGRAM(I)=NGRAMX                                                      NITR
       CRNO(I)=CRNOL                                                       DEC
      NFNH3(I)=NFNH3L                                                      DEC
        NGRAM(I+4)=NRNL                                                    DEC
         CNL(I)=CNCML                                                      DEC
      TLRCO2=TLRCO2+RLCO2L                                                 DEC
   22 CONTINUE                                                             DEC

*********************************************************************

*   CALCULATION OF NET FLOWS

*  NFNO= NET FLOW OF NO3 ; FOR THE FIRST 4 LAYERS DIFFERENT
*  FROM 5TH-20TH LAYER, MG N PER DAY

      DO 23  I=1,4                                                         SALT
      NFNO(I)=CRNO(I)+RNITF(I)+SFLW(I)-SFLW(I+1)                           SALT
   23 CONTINUE                                                             SALT

      DO 24 I=5,N                                                          NITR
      NFNO(I)=SFLW(I)-SFLW(I+1)                                            NITR
   24 CONTINUE                                                             NITR

      DO  25  I=1,N                                                        WAT
      NFLW(I)=FLRW(I)-FLRW(I+1)                                            WAT
*  NFLW = NET FLOW OF WATER, CM/DAY
   25 CONTINUE                                                             WAT

      DO  26  I=1,10                                                       TEMP
      NHFL(I)=HEFLW(I)-HEFLW(I+1)+HEDI(I)-HEDI(I+1)                        TEMP
*  NET FLOW OF HEAT, CAL/DAY*CM**2
   26 CONTINUE                                                             TEMP
```

```
*   CALCULATION OF AMOUNTS AND  VOLUMES BY INTEGRAL FUNCTIONS

*   VOLUME WATER PER LAYER, CM**3
      VOLW1=INTGRL(IVLW1,NFLW1,20)                                   WAT

*   AMOUNT OF NO3 PER LAYER, MG N
      NO3B1=INTGRL(INOB1,NFNO1,20)                                   SALT

*     HEATH CAPACITY PER LAYER,   CAL
      HCL1=INTGRL(IHCL1,NHFL1,10)                                    TEMP

*    COMPONENTS OF ORGANIC MATTER,   MG  C PER LAYER

      COMP1 =INTGRL(ICMP1 ,RDC1 , 4  )                               DEC
      COMP5 =INTGRL(ICMP5 ,RDC5 ,4 )                                 DEC
      COMP9 =INTGRL(ICMP9 ,RDC9 ,4 )                                 DEC
      COMP13=INTGRL(ICMP13,RDC13,4 )                                 DEC
      COMP17=INTGRL(ICMP17,RDC17,4 )                                 DEC

*    AMOUNT OF AMMONIUM PER LAYER, MG N
       NH3B1=INTGRL(INH3B1,NFNH31,4)                                 NITR


*    AMOUNT OF CO2, MG C FOR ALL LAYERS
       BCO2=INTGRL(10.,TLRCO2)                                       DEC


*    NUMBER OF AMMONIUM OXIDIZERS PER LAYER
       AMOX1=INTGRL(IAMOX1,NGRAM1,4)                                 NITR


*    AMOUNT OF ORGANIC NITROGEN IN LIGNIN PER LAYER
          AMOX5=INTGRL(IAMOX5,NGRAM5,4)                              DEC

*         H=HYDRAULIC HEAD OF WATER, CM
          H=INTGRL(0.,IH)                                            WAT

*      OUTFLOW OF SALT OVER LAST BOUNDERY, MG N
       OUTFLS=INTGRL(10.,SFLW(N+1))                                  SALT



*         INFLW=INFLOW OF WATER BY IRRIGATION, CM
             INFLWW= INTGRL(10.,FLRW(1))                             WAT


*   OUTFLOW  OF WATER OVER LAST BOUNDERY, CM
             OUTFLW = INTGRL(10.,FLRW(N+1))                          WAT

*************************************************************************

*   CONVERTING THE OUTPUT FROM MG  AND CM**3 TO RESP. KG AND  M**3
*   PER HECTARE PER LAYER THICKNESS
*           FAC2 IS  CONVERSION FACTOR
PARAMETER              FAC2=1.E2                                     CONV

         INFLW=(INFLWW-10.)*FAC2                                     CONV
      OUTFL=(OUTFLW-10.)*FAC2                                        CONV
      OUTFS=(OUTFLS-10.)*FAC2                                        CONV
      BCO2HA=(BCO2-10.)*FAC2                                         CONV
```

```
* HNL IS AMOUNT OF NITROGEN IN LIGNIN

      DO  27  I=1,4                                                      CONV
      HNL(I)=AMOX(I+4) *FAC2                                             CONV
      PCOMP(I)=COMP(I)*FAC2                                              CONV
      SCOMP(I)=COMP(I+4 )*FAC2                                           CONV
      LCOMP(I)=COMP(I+8 )*FAC2                                           CONV
      CCOMP(I)=COMP(I+12)*FAC2                                           CONV
      BCOMP(I)=COMP(I+16)*FAC2                                           CONV
      TOTCOM(I)=PCOMP(I)+SCOMP(I)+LCOMP(I)+CCOMP(I)                      CONV
   27 AMMON(I)=NH3B(I)*FAC2                                              CONV

* TOTCOM IS SUM OF THE COMPONENTS OF ORG MATTER PER LAYER PER HA

      DO 28 I=1,20                                                       CONV
   28 NITRAT(I)=NO3B(I)*FAC2                                             CONV


      TORGM1=TOTCOM(1)                                                   OUT
      TORGM4=TOTCOM(4)                                                   OUT
      WCGM1 = WC(1)
      NITRA1 = NITRAT(1)
      NITRA4 = NITRAT(4)
      AMMON1 = AMMON(1)
      AMMON4 = AMMON(4)
      CNLGM1 = CNL(1)
AMOXGM = AMOX(1)
TGM1 = T(1)
************************************************************************

METHOD MILNE

TIMER FINTIM = 45., PRDEL =2., OUTDEL = 1., DELT = 0.005
FINISH TELLER = 13000.
PRINT OUTFL,OUTFS,BCO2HA,TELLER,H,INFLW                                  OUT
LABEL     TOTAL AMOUNT OF ORG MATTER OF FIRST LAYER PER HA               OUT
PRTPLOT   TORGM1(0.,)                                                    OUT
LABEL     TOTAL AMOUNT OF ORG MATTER OF FOURTH LAYER PER HA              OUT
PRTPLOT   TORGM4 (0.,)                                                   OUT

PRTPLOT   WCGM1     (0.,)
PRTPLOT   NITRA1    (0.,)
PRTPLOT   NITRA4    (0.,)
PRTPLOT   AMMON1    (0.,)
PRTPLOT   AMMON4    (0.,)
PRTPLOT   CNLGM1    (0.,)
PRTPLOT   AMOXGM    (0.,)
PRTPLOT   TELLER    (0.,)
*                                  DEBUG SECTION NOT IN USE
*     IF(KEEP.LT.0.5) GOTO 29                                            OUT
*     Y=DEBUG(5,0.)                                                      OUT

   29 CONTINUE                                                           OUT
      A=IMPULS(0.,PRDEL)                                                 OUT
      IF(A*KEEP.LT.0.5)  GOTO 30                                         OUT

* OUTPUT

WRITE (6, 102)
WRITE (6, 102)
WRITE (6, 102)
WRITE (6, 102)
102   FORMAT(1H )

            WRITE(6,81)(WC(I),I=1,N)                                     OUT
   81 FORMAT(1H ,34HWATER CONTENT FOR DIFFERENT DEPTHS//(10F10.4/))      OUT

      WRITE(6,82) (T(I),I=1,10)
82    FORMAT(1H ,26HTEMPERATURES OF THE LAYERS//(10F10.2/))
```

```
      WRITE(6,83) (CNO(I),I=1,N)                                                OUT
 83   FORMAT(1H ,38HCONCENTRATIONS NO3 PER LAYER MGN/CM**3//(10F10.6/)) OUT

      WRITE(6,84)(PCCMP(I),SCCMP(I),LCCMP(I),CCCMP(I),BCCMP(I),I=1,4)   OUT
 84   FORMAT(1H ,74HPROTEINS         SUGARS      LIGNIN      CELLULOSE   BIOUT
     /OMASS KGC/HA PER LAYER//(5F12.3/))                                         OUT

      WRITE(6,85)(NITRAT(I),I=1,N)                                              OUT
 85   FORMAT(1H ,29HNITRATE KG N PER HA PER LAYER//(10F10.3/))                  OUT

      WRITE(6,86) (AMMON(I),I=1,4)                                              OUT
 86   FORMAT(1H ,28HAMMON  KG N PER HA PER LAYER//(4F10.3))                     OUT
WRITE (6, 102)

      WRITE(6,87) (CNL(I),I=1,4)                                                OUT
 87   FORMAT(1H ,30HC/N-RATIO OF LIGNIN  PER LAYER//(4F10.1))                   OUT
WRITE (6, 102)

      WRITE(6,88) (HNL(I),I=1,4)                                                OUT
 88   FORMAT(1H ,40HHNL  KG ORGANIC N IN LIGNIN PER HA*LAYER//(4F10.2)) OUT
WRITE (6, 102)

      WRITE(6,89) (AMOX(I),I=1,4)                                               OUT
 89   FORMAT(1H ,22HAMOX PER LAYER PER CM2//(4F10.1))                           OUT
WRITE (6, 102)

 30   CONTINUE
END

STOP
```

UNIV. COLL. N.W.
BANGOR
LIBRARY